Digitalization and Development

This book examines the diffusion of digitalization and Industry 4.0 technologies in Malaysia by focusing on the ecosystem critical for its expansion. The chapters examine the digital proliferation in major sectors of agriculture, manufacturing, e-commerce and services, as well as the intermediary organizations essential for the orderly performance of socioeconomic agents.

The book incisively reviews policy instruments critical for the effective and orderly development of the embedding organizations, and the regulatory framework needed to quicken the appropriation of socioeconomic synergies from digitalization and Industry 4.0 technologies. It highlights the importance of collaboration between government, academic and industry partners, as well as makes key recommendations on how to encourage adoption of IR4.0 technologies in the short- and long-term.

This book bridges the concepts and applications of digitalization and Industry 4.0 and will be a must-read for policy makers seeking to quicken the adoption of its technologies.

Rajah Rasiah is Distinguished Professor of Economics and Executive Director of the Asia-Europe Institute, Universiti Malaya.

Wah Yun Low is Honorary Professor of Psychology at the Faculty of Medicine, Universiti Malaya.

Nurliana Kamaruddin is Deputy Executive Director of Academic and Students Affairs and Senior Lecturer at the Asia-Europe Institute, Universiti Malaya.

Routledge Frontiers of Business Management

Innovation and Industrial Development in China
A Schumpeterian Perspective on China's Economic Transformation
Kaidong Feng

Beauty of Crowdfunding
Blooming Creativity and Innovation in the Digital Era
Sunghan Ryu

Renewable Energy Management in Emerging Economies
Strategies for Growth
Henry K. H. Wang

Statistical Modelling and Sports Business Analytics
Edited by Vanessa Ratten and Ted Hayduk

China's Drive for the Technology Frontier
Indigenous Innovation in the High-Tech Industry
Yin Li

Digital Transformation in Aviation, Tourism and Hospitality in Southeast Asia
Edited by Azizul Hassan and Nor Aida Abdul Rahman

Football Entrepreneurship
Edited by Vanessa Ratten

Digitalization and Development
Ecosystem for Promoting Industrial Revolution 4.0 Technologies in Malaysia
Edited by Rajah Rasiah, Wah Yun Low, and Nurliana Kamaruddin

For more information about this series, please visit www.routledge.com/ Routledge-Frontiers-of-Business-Management/book-series/RFBM

Digitalization and Development

Ecosystem for Promoting Industrial
Revolution 4.0 Technologies in Malaysia

**Edited by
Rajah Rasiah, Wah Yun Low,
and Nurliana Kamaruddin**

Routledge
Taylor & Francis Group

LONDON AND NEW YORK

First published 2023
by Routledge
4 Park Square, Milton Park, Abingdon, Oxon OX14 4RN

and by Routledge
605 Third Avenue, New York, NY 10158

Routledge is an imprint of the Taylor & Francis Group, an informa business

British Library Cataloguing-in-Publication Data
A catalogue record for this book is available from the British Library

Library of Congress Cataloging-in-Publication Data
Names: Rasiah, Rajah, editor. | Low, Wah Yun, editor. | Nurliana Kamaruddin, editor.
Title: Digitalization and development : ecosystem for promoting industrial revolution 4.0 technologies in Malaysia / edited by Rajah Rasiah, Wah Yun Low, and Nurliana Kamaruddin.
Description: New York, NY : Routledge, 2023. |
Series: Routledge frontiers of business management | Includes bibliographical references and index. |
Identifiers: LCCN 2022046528 (print) | LCCN 2022046529 (ebook) | ISBN 9781032433066 (hardback) | ISBN 9781032433950 (paperback) | ISBN 9781003367093 (ebook)
Subjects: LCSH: Manufacturing industries—Technological innovations—Malaysia. | Computer integrated manufacturing systems—Malaysia. | Production management—Data processing. | Technological innovations—Management—Malaysia.
Classification: LCC HD9736.M35 D54 2023 (print) | LCC HD9736.M35 (ebook) | DDC 338.4/76709595—dc23/eng/20221118
LC record available at https://lccn.loc.gov/2022046528
LC ebook record available at https://lccn.loc.gov/2022046529

ISBN: 9781032433066 (hbk)
ISBN: 9781032433950 (pbk)
ISBN: 9781003367093 (ebk)

DOI: 10.4324/9781003367093

Typeset in Galliard
by codeMantra

Contents

Figures

Tables

Contributors

Norlidah Alias is currently Associate Professor in the Curriculum and Instructional Technology Department, Faculty of Education, Universiti Malaya. She obtained her doctorate from the University of Malaya and has been involved in research projects related to curriculum development, digital pedagogies, TVET, and Environmental Education. She is a principal consultant of projects related to 21st-century learning and curriculum development and has acted as an assessor for the Turkish Higher Education Quality Council.

Cheong Kee Cheok is Adjunct Professor at the Faculty of Business and Economics and simultaneously Senior Advisor at the Asia-Europe Institute, Universiti Malaya. With a PhD from the London School of Economics, he was staff of the World Bank for nearly 20 years before returning to his alma mater, Universiti Malaya to take up a senior research fellowship. He has been Adjunct Professor since 2020.

Maslina Daud is currently the Senior Vice-President of Cybersecurity Proactive Services division at CyberSecurity Malaysia. She obtained her doctorate in cybersecurity from University of Malaya. Her areas of specialization include information security, cybersecurity, and data privacy.

Dorothy DeWitt is currently Associate Professor in the Curriculum and Instructional Technology Department, Faculty of Education, Universiti Malaya. She obtained her doctorate from the University of Malaya and was a recipient of the Endeavour Executive Fellowship at Macquarie University. She has also won awards for teaching innovations at the national and international levels, which include Best Immersive Learning Showcase at iLRN2020, the Universiti Malaya Excellence Award for Innovative E-Learning in 2018, and Runners-Up for Minister of Education's Malaysia Special Awards (Immersive Teaching and Learning) in 2019.

Fumitaka Furuoka is Associate Professor at Asia-Europe Institute, Universiti Malaya. His research interests include labour economics and international economics.

Azirah Hashim is Emeritus Professor at the Faculty of Languages and Linguistics (FLL), Universiti Malaya. She has held previously several positions, including the Executive Director of the Asia-Europe Institute and Dean of FLL. Currently, she is President of the International Association of Applied Linguistics (AILA) and Ambassador Scientist for the Alexander von Humboldt Foundation, Germany.

Nurliana Kamaruddin is currently the Deputy Executive Director of Academic and Students Affairs and Senior Lecturer at the Asia-Europe Institute, Universiti Malaya. She obtained her PhD in Development Cooperation from Ewha Womans University, South Korea and was a recipient of the Korea Foundation ASEAN Fellowship, the POSCO TJ Park Foundation Asia Fellowship (2009–2011), and the Ewha Global Partnership Program (2006–2009).

Kiranjeet Kaur is currently a PhD candidate at the Asia-Europe Institute, Universiti Malaya. Her research revolves around the area of advance technology adoption in the textiles and clothing manufacturing industry in Malaysia with a focus on drawing policy implications to quicken the diffusion of Industry 4.0 technologies in the industry. She is a finance professional specializing on strategy, product development, and marketing with an emphasis on digitalization.

Hemant Kedia is an engineer with a postgraduate degree in business management. He has over 25 years of experience across different countries with profound know-how in techno-commercial areas associated with operations management, project management, cost management, mergers and acquisitions, and post-acquisition strategic integration, sourcing, and business development. He is presently the head of sourcing, strategy and business development at Recron, which is the largest integrated polyester and textile company in Malaysia.

Abdul Latif is currently Director General, Malaysia Productivity Corporation (MPC), Ministry of International Trade and Industry (MITI). He is also the Chairperson of Delivery Management Office MPC, and oversees the implementation of initiatives by all the 11 Sectoral Productivity Nexus under the Malaysia Productivity Blueprint (MPB) launched by YAB Prime Minister on May 2017. He holds a Master's degree in Policy Studies from Saitama University. Latif also leads the Secretariat of the Special Taskforce to facilitate Business (PEMUDAH) and had also attended the Advanced Management Program at INSEAD, Fontainebleau, France.

Yeap Khai Leang is a PhD candidate at the Asia-Europe Institute, Universiti Malaya. Her research examines the diffusion of Industry 4.0 in Malaysia's electronics manufacturing industry. She had previously served as a graduate research assistant under the supervision of Professor Rajah Rasiah, funded by the *Tabung Khas Profesor Ulung*. She is currently Economic Development Adviser at the UK Foreign Commonwealth and Development Office, the British High Commission in Kuala Lumpur.

Jingyi Li is a PhD candidate at the Asia-Europe Institute, Universiti Malaya. Her research interests include COVID-19 and tourism studies.

Beatrice Lim is Senior Lecturer at the Faculty of Business, Economics and Accountancy, Universiti Malaysia Sabah. Her research interests include labour economics and gender studies.

Wah Yun Low is Honorary Professor of Psychology at the Faculty of Medicine, Universiti Malaya, Kuala Lumpur. She graduated with a PhD from the University of Surrey, England. She was the former Deputy Executive Director of Research & Internationalisation, Asia-Europe Institute, Universiti Malaya. She was the Immediate Past President, Asia-Pacific Academic Consortium for Public Health. She is the Editor-in-Chief of the *Asia-Pacific Journal of Public Health*.

Cecilia Cheong Yin Mei is affiliated to the Faculty of Languages and Linguistics, Universiti Malaya. She was formerly the Faculty's Head of Multimedia Planning Unit and was involved in an international joint research project on the meaning of quality in online blended teaching and learning at the tertiary level. She is currently the Faculty Coordinator of open and distance learning programmes.

Yip Tien Ming is a PhD candidate at the Faculty of Economics and Business, Universiti Malaya. His research interests include applied econometric methods and demography.

Shankaran Nambiar is Senior Research Fellow at the Malaysian Institute of Economic Research. He is currently Visiting Fellow at the Crawford School of Public Policy, Australian National University.

Nazreen Mohd Nasir is currently Senior Consultant at Strand Aerospace Malaysia. He obtained his doctorate in Engineering from the International Islamic University, Malaysia (IIUM). His recent work is in strategic road mapping towards industrial transformation and human capital development with both the public and private sectors. He has been involved in several OEM aircraft certification programmes and turnkey projects on IR4.0 technologies. He has also served as industrial mentor for several human capital programmes.

Naguib Mohd Nor is currently Chief Executive Officer of Strand Aerospace Malaysia and President of Malaysia Aerospace Industry Association (MAIA). He has been active engineer, technologist, and business developer in the global aerospace supply chain since 2000 and speaks frequently on aerospace and other technology subjects at global events. He also works closely with the government on Industry 4.0 strategy and implementation. He holds a BEng in Aerospace Engineering from UMIST and a MSc in Aerospace Vehicle Design from Cranfield.

Suresh Palpanaban is the Strategy Leader responsible for the design and delivery of DreamEDGE's innovations, product, and engineering solutions in the fields of artificial intelligence, augmented reality, virtual reality, automation, and electric mobility leveraging on internal capability for the verticals of automotive, aerospace, maritime, rail, and electric vehicles.

Khairul Hanim Pazim is Senior Lecturer at the Faculty of Business, Economics and Accountancy, Universiti Malaysia Sabah. Her research interests include labour economics and aging studies.

Li Ran is Senior Lecturer at the Institute of China Studies, Universiti Malaya. She obtained her doctoral degree in Economics from Universiti Malaya in 2014. Her specialization is in the transformation of China's state enterprises, state enterprise system, and China's political-economic system. Her current research focuses on China's global strategy and China-Malaysia economic relations.

Mohd Afzanizam Abdul Rashid is former Chief Economist at Bank Islam Malaysia Berhad and is currently associated with the Employees Provident Fund as Head of Economics and Research Department. He obtained his doctorate in Economics from Universiti Kebangsaan Malaysia. He is also Adjunct Professor at Faculty of Business and Management, Universiti Teknologi MARA.

Rajah Rasiah is currently Distinguished Professor of Economics and Executive Director of the Asia-Europe Institute, Universiti Malaya. He obtained his doctorate in Economics from Cambridge and was a fellow at Harvard University. He is the recipient of the Celso Furtado Prize from the World Academy of Sciences for advancing the frontiers of social science thought (Economics) in 2015 and Malaysia's Merdeka Prize for academic excellence in 2018.

Saliza Saari is currently Director of the Malaysia Productivity Corporation (MPC). She holds a Master's degree in Knowledge Management from University Technology MARA. She specializes in productivity and quality improvement consultancy projects and is certified as a small and medium enterprise Business Counsellor by Japan-SMIDEC. She is also Lead Assessor for the Prime Minister's Award, Malaysia.

Yosuke Uchiyama is a PhD candidate in the Department of East Asian Studies, Faculty of Arts and Social Sciences, Universiti Malaya. His research interests include gig economy and sharing economy.

Tham Siew Yean is Emeritus Professor at Universiti Kebangsaan Malaysia. She is currently Visiting Senior Fellow at ISEAS-Yusof Ishak Institute, Singapore. She is also Associate Fellow at Institute of China Studies, University of Malaya and External Fellow at Jeffery Cheah Institute on Southeast Asia, Sunway University. She obtained her PhD from University of Rochester.

Andrew Kam Jia Yi is Associate Professor/Senior Research Fellow at the Institute of Malaysian and International Studies (IKMAS), Universiti Kebangsaan Malaysia. He received the Chevening scholarship for his MSc in Economics at the University of Warwick, UK, and his PhD at the Australian National University under the Endeavour Postgraduate Scholarship. He was also a Fulbright Scholar at the University of California Santa Barbara (2015/2016). He is also a consultant to EPU, MITI, MDEC, UNCTAD, and UNESCO.

Foreword

The importance of digitalization was amplified during the COVID-19 pandemic, accelerating the digital transformation process for many nations. The Fourth Industrial Revolution (IR4.0) led this process, facilitating the adoption of new technologies such as artificial intelligence (AI), cloud, and big data in all aspects of our lives.

The Malaysian government launched the National IR4.0 Policy on July 1, 2021, with the aim of ensuring the country is equipped to benefit from the tide of IR4.0. The National IR4.0 Policy consists of mainly four policy thrusts, namely, equipping the rakyat with the requisite knowledge and skill sets; forging a connected nation through digital infrastructure development; future-proofing regulations to be compatible with technological changes; and accelerating IR4.0 technology innovation and adoption.

As this book highlights, it is also important to scrutinize embedding ecosystem pillars to ensure the implementation of IR4.0 technologies in the country is steady, smooth, and can achieve the desired results.

The expansion of IR4.0 is especially crucial for the service industry including manufacturing, agriculture, and selected organizations. These are some of the sectors that are undergoing a radical transformation. This also brings to the fore the importance of cybersecurity measures to protect these sectors. I always say that cybersecurity is the immune system in the digitalization and operations of any ICT sector.

What is important to note is that IR4.0 cannot be achieved in isolation. It must involve collaborative efforts of the government, academia, and industry. Leveraging each of our strengths, we can create conducive ecosystems to accelerate digitalization via the convergence of technologies, disciplines, talents, and by facilitating the emergence of new business models.

The research process to formulate this book has provided us with the tools to take stock of past successes as well as failures and to chart the path forward in light of the challenges of today. Fluidity and agility are important factors in today's environment and Huawei Malaysia will be the agent of change to assist in publicizing the strategies and input recommended by the authors of this book for our future development as well as in being the driver of intellectual, social, and economic change to ensure Malaysia is at the forefront of digital transformation and innovation.

This book also represents the concerted efforts by Huawei's stakeholders to see the industry ecosystem speed up its efforts to harness scientific and technological achievements to boost productivity, unleash new impetus for innovation-driven growth, enhance Malaysia's capacity for development, and contribute to its rapid development. We shall continue to engage in advanced research, especially in areas globally recognized as our strength, and provide an array of digital solutions such as our technology in connectivity, artificial intelligence, and cloud services.

We aim to enrich the learning experience of those we partner with and enhance competencies to contribute to a globally competitive and rapidly changing environment. To implement these strategic aims, we need to develop an enabling ecosystem in which our resources are appropriately allocated and deployed to help us attain sustainable excellence.

With these in place, we are sure the goals to transform Malaysia into a digitally driven, high-income nation and a regional leader in ASEAN will be realized. By investing in the future of technology and the digital economy, we are investing in a better future for Malaysia and its communities.

Congratulations to the authors of the book, *Digitalisation and Development - Ecosystem for Promoting Industrial Revolution 4.0 Technologies in Malaysia*, which outlines the strategic visions and goals to help Malaysia realize its full potential. The publication of this book has also provided us with some food for thought that will help us in ensuring the success of IR4.0 in Malaysia. Progress can only be achieved through cooperation in a fully connected Malaysia and with our partners, we aim to achieve exactly that.

Michael Yuan
Chairman, Huawei Malaysia

Preface

The Fourth Industrial Revolution (IR4.0) distinguishes itself from the simple automation that characterized IR3.0 when the implementation of artificial intelligence emerged to support human activity. Autonomous and unmanned robots and drones have become increasingly important in augmenting human activity, so much so that governments across the world have strategized to quicken its diffusion through proactive promotion and the construction of critical infrastructure to facilitate the process such as the installation of broadband cables for the spread of big data analytics, internet of things, and cloud computing. Indeed, although digitalization has been actively promoted in Malaysia since 2010, the government only launched the IR4.0 Master Plan in 2018 and the digitalization blueprint in 2020.

The spread of IR4.0 technologies has seen the use of robots and drones to plough fields; harvest fruits; water, fertilize, and spray insecticide for plants; offer cyber-links between experimental locations and researchers; support autonomous driverless vehicles; load and unload materials onto vehicles; and to connect and coordinate activities of humans through contactless operations. The displacement of human labour by IR4.0 instruments has helped some countries (such as Singapore, Taiwan and Japan) overcome labour shortage problems, but threatens to raise unemployment levels in labour surplus countries such as Bangladesh, Indonesia, and Pakistan. At the broader philosophical level, the spread of such disruptive technologies has attracted serious debate concerning digital equalization and the digital divide. Consequently, there is an understanding that unless empowerment and alternative sources of equalizing support are enjoyed by all members of society and unless socioeconomic agents can catch the waves of change necessitated by the spread of IR4.0 technologies, the appropriation of IR4.0 synergies will be uneven.

This book examines the elements of public goods associated with upgrading the embedding environment to ensure that the introduction and spread of IR4.0 technologies in Malaysia are promoted smoothly to raise the economy, as well as ensure that learning and security synergies are addressed effectively. Since no one single book can exhaustively address the myriad of issues associated with the diffusion of digitalization and IR4.0 technologies, this book covers some critical embedding ecosystem pillars that need scrutiny in Malaysia, which are presented through 13 chapters. It focuses on the three major sectors of services, manufacturing, and agriculture, and the intermediary role of selected organizations, and cybersecurity measures to manage the safe expansion and operations of IR4.0 instruments.

We take this opportunity to acknowledge several members without whose support the preparation of this book would not have been possible. Foremost on the list is the management of Huawei Corporation, which offered us a generous grant to support the research behind the chapters assembled for the book. In doing so, we thank Michael Yang Ming, chief executive officer, and Oliver Lu Jin-Feng from Huawei Technologies (Malaysia) for their pivotal support for this book project. Next, we would like to acknowledge financial support from the Distinguished Professor of Economics grant awarded by the Ministry of Higher Education, Malaysia to Rajah Rasiah in 2018 (Project no.: MO-004–2018), which in addition helped finance the fieldwork undertaken by Yeap Khai Leang and Kiranjeet Kaur. We wish to also acknowledge the support from the Malaysian Productivity Corporation (MPC), the Malaysian Institute of Economic Research (MIER), Strand Aerospace Malaysia (STRAND), and DreamEdge for their participation in the initial workshop to launch the book project. We wish to thank Associate Professor Dr Jatswan Singh Sidhu, the former executive director of the Asia-Europe Institute, for his support and encouragement.

Rajah Rasiah, Wah Yun Low, and Nurliana Kamaruddin

List of Abbreviations

3D	Three-Dimensional
4G	Fourth-Generation Wireless
5G	Fifth-Generation Wireless
5G	Fifth-Generation Technology
11MP	Eleventh Malaysia Plan
12MP	Twelfth Malaysia Plan
ABA	Air Business Academy
AFPN	Agro-food Productivity Nexus
AI	Artificial Intelligence
AMIC	Aerospace Malaysia Innovation Centre
API	Application Programming Interface
APRU	Average Revenue Per User
ASEAN	Association of Southeast Asian Nations
B2B	Business-to-Business
B2C	Business-to-Consumer
B2G	Business-to-Government
BAU	Business as Usual
BBC	British Broadcasting Corporation
BCM	Business Continuity Management
BNM	Bank Negara Malaysia
CAAM	Civil Aviation Authority of Malaysia
CAD	Computer-Aided design
CGC	Credit Guarantee Corporations
CIMB	Commerce International Merchant Bank
CK	Content Knowledge
COE	Centre of ExcellenceCOVID-19 Coronavirus Disease
CPN	Chemical Productivity Nexus
CPS	Cyber-Physical Systems
CPPS	Cyber-Physical Production Systems
CPTPP	Comprehensive and Progressive Agreement for Trans-Pacific Partnership
CSA	Community-Supported Agriculture
CTRM	Composites Technology Research Malaysia

DCA	Department of Civil Aviation
DFTZ	Digital Free Trade Zone
DNB	Digital Nasional Berhad
DOSM	Department of Statistics Malaysia
DPN	Digital Productivity Nexus
DT	Digital Trade
DTD	Digital Talent Development
DTDA	Digital Trade Restrictiveness Index
DTMS	Digital Talent Maturity Score
DTT	Digital Technology Transformation
EBIC	Electronic Beam-Induced Current
EBOP	Extended Balance of Payments
ECER	East Coast Economic Region
E&E	Electric and Electronics
EEPN	Electric and Electronics Productivity Nexus
EPU	Economic Planning Unit
E-SWOT	Electronic-Strengths, Weaknesses, Opportunities and Threats
FDI	Foreign Direct Investment
FELDA	Federal Land Development Authority
FMCG	Fast-Moving Consumer Goods
FMM	Federation of Malaysian Manufacturers
FTA	Free Trade Agreement
GCI	Global Cybersecurity Index
GDP	Gross Domestic Product
GLIC	Government-Linked Investment Company
GMM	Generalized Method of moments
GMV	Gross Merchandise Value
HRD	Human Resource Development
HRDB	Human Resource Development Board
HRDF	Human Resource Development Fund
IB	Internet Banking Penetration Rate
ICS	Industrial Control System
ICT	Information Communication Technology
IIH	International Innovation Hub
IIoT	Industrial Internet of Thing
ILO	International Labour Organization
IM	Iskandar Malaysia
IoS	Internet of Services
IoT	Internet of Things
IPI	Industrial Production Index
IR	Industrial Revolution
IR3.0	Third Industrial Revolution
IR4.0	Fourth Industrial Revolution
ISMS	Information Security Management System
ISO	International Standard Organization

ISP	Internet Service Providers
ISTE	International Society for Technology in Education
IT	Information Technology
JSI	Joint Statement Initiative
KKMM	Ministry of Communications and Multimedia
LDC	Least Developed Country
M2M	Machine-to-Machine Communication
MADA	Muda Agriculture Development Authority
MAFI	Ministry of Agriculture and Food Industries
MARDI	Malaysian Agriculture Research and Development Institute
MATA	Malaysia Automation Technology Association
MATRADE	Malaysia External Trade Development Corporation
MB	Megabyte
MCMC	Malaysia Communications and Multimedia Commission
MCO	Movement Control Order
MCSS	Malaysia Cyber Security Strategy
MDB	Malaysian Digital Blueprint
MDEC	Malaysia Digital Economy Corporation
MEPN	Machinery and Equipment Productivity Nexus
MEIR	Malaysian Institute of Economic Research
MEVAC	Machinery and Equipment Virtual Advisory Centre
MIDA	Malaysian Industrial Development Authority
Misi4.0	Malaysia Industry 4.0 Systems Integrator Association
MITI	Ministry of International Trade and Industry
MKMA	Malaysian Knitting Manufacturing Association
MOBILE	Mobile banking penetration rate
MOOC	Massive Open Online Course
MPB	Malaysia Productivity Blueprint
MSIC	Malaysia Standard Industrial Classification
MSME	Small and Medium Enterprise
MTMA	Malaysian Textiles Manufacturing Association
MYRUSD	Dollar to Ringgit Exchange Rate
NCER	Northern Corridor Economic Region
NESR	National E-commerce Strategic Roadmap
NITA	National Information Technology Agenda
OECD	Organisation for Economic Co-operation and Development
OEM	Original Equipment Manufacturer
OFD	Online Food Delivery
OT	Operational Technology
PENJANA	*Pelan Jana Semula Ekonomi Negara*
PIP	Productivity Improvement Programme
PK	Pedagogical Knowledge
PWD	Person with Disabilities
R&D	Research and Development
RCEP	Regional Comprehensive Economic Partnership

RFBPN	Retail and Food and Beverage Productivity Nexus
RFBVAC	Retail and Food and Beverage Virtual Advisory Centre
RPA	Robotic Process Automation
RR	Rolls Royce
RWA	Risk Weighted Asset
SCADA	Supervisory Control and Data Acquisition
SCRE	Sarawak Corridor of Renewable Energy
SDC	Sabah Development Corridor
SDG	Sustainable Development Goal
SIM	Stratified Investment Model
SME	Small and Medium Enterprise
SOP	Standard Operating Procedure
STI	Science, Technology, and Innovation
STRAND	Strand Aerospace Malaysia
THOU	Tun Hussein Onn University
TK	Technological Knowledge
TOE	Technology-Organization-Environment
TPACK	Technological Pedagogical Content knowledge
TPM	Technology Park Malaysia
TTRI	Taiwan Textile Research Institute
TVET	Technical and Vocational Education and Training
UAV	Unmanned Aerial Vehicle
UiTM	Universiti Teknologi Mara
UMW	United Motor Works
UN	United Nations
UNCTAD	United National Conference on Trade and Development
UNESCO	United Nations Educational, Scientific and Cultural Organization
VAPT	Vulnerability Assessment and Penetration Testing
VoIP	Voice over Internet Protocol
WEF	World Economic Forum
WIPO	World Intellectual Property Organization
WTO	World Trade Organization

1 Problematizing Digitalization and Industrial Revolution 4.0

Rajah Rasiah, Nurliana Kamaruddin, and Wah Yun Low

Introduction

While the proliferation of digitalization started from the 1990s when broadband cables using fibre optics were laid in several parts of the world (Arai, Naganuma & Satake, 2012; Arai, 2019), it also provided the foundation for the emergence and expansion of Industry 4.0 technologies. Scattered developments taking advantage of the emerging digital infrastructure was happening around the turn of the millennium. For example, BMW had its assembly of right-hand steering wheel cars in South Africa controlled from its headquarters in Germany by 2002 to meet its just-in-time delivery inventory control system (Rasiah, 2004). By 2010, Fourth Industrial Revolution (IR4.0) technologies had emerged as a major area of focus on government policy blueprints targeted at quickening structural transformation with formal application of the epochal shift from Third Industrial Revolution (IR3.0) to IR4.0 recognized in Germany in 2011. Consequently, countries began expanding the laying of broadband infrastructure, internet of things, big data analytics, and cloud computing facilities to stimulate the use of robots and drones in economic activity. While specialized firms expanded in numbers to develop robots and drones for a wide range of purposes, governments began developing the infrastructure (or ecosystem) to stimulate their use.

Whereas the emergence of IR4.0 technologies has driven productive changes in production, distribution, and consumption methods, it has also awakened an old debate on new technologies (United Nations, 2021). Each time the world faces epochal changes from the advent of new technologies, the ensuing debates has continued to capture two sides of coin arguments, *viz.*, on the one side, glorifying the potential such technologies bring, and on the other, carrying the potential doom such technologies wreak on humanity. The mystical figure Ned Ludd emerged when workers rebelled in 1799 fearing that mechanization would drive out skilled jobs. Highly skilful weavers and textile workers then had objected to the introduction of mechanized looms and knitting machines, and consequently, protested by breaking machines and burning factories after the British government did not respond to their concerns. However, the Luddite resistance to the introduction of mechanization ended by 1813 (Andrews, 2019). Especially, manufacturing absorbed the key innovations of standardization,

DOI: 10.4324/9781003367093-1

inter-changeability, and flow (conveyor belt) to stimulate mass production (Best, 1990, 2001, 2018). Jobs lost from technological change were quickly created in other activities as the concerns over mass retrenchments and unemployment did not materialize. While deskilling and the restructuring of jobs that reduced worker tasks to highly differentiated, monotonous, and repetitive activities that robbed some jobs of creative work, it also created hierarchies that supported management and technical tasks (Taylor, 1911), which was later differentiated further under Fordism despite a rise in technical specialization and wages before being creatively flexibilized under neo-Fordism (Tolliday & Zeitlin, 1987).[1]

Industries in which technical change was slow and product cycles long, such as clothing, locations enjoying low wages and large amounts of surplus labour (e.g., Bangladesh and Pakistan) faced little transformation from labour intensive operations to the deployment of cutting-edge technologies. A handful of firms in these countries introduced state-of-the-art technologies only because of the precision, quality, and defect-less processing that the technologies allowed, which again was largely conditioned by buyers (Nazia & Rasiah, 2022). In industries where technical change has been swift and where production has evolved to be too sophisticated for the use of manual labour, such as highly miniaturized semiconductors, only locations with technically sound labour and modern infrastructure have managed to retain such operations. In fact, the lack of research universities engaged in frontier R&D activities has inhibited catch-up into frontier wafer fabrication activities in Singapore, Malaysia, and Thailand (Rasiah, 2020).

Digitalization and IR4.0 technologies have penetrated all walks of life, which is why it has become an integral component of the 17 sustainable development goals launched by the United Nations (United Nations, 2015). These technologies are transforming all sectors, including agriculture, education, and services. While the COVID-19 pandemic has quickened its introduction in some countries to contain the spread of the virus through social distancing to prevent the spread of the infectious disease, countries, such as Israel and Taiwan embarked on it earlier to appropriate economic synergies from such technologies. Taiwan, for example, has relocated back a number of its manufacturing firms and increased its self-sufficiency through agricultural programmes by increasing the deployment of robots and drones (Rasiah, 2007, 2019). Huawei (2021) presents some of the exciting developments on what IR4.0 technologies offer humankind for the near future.

As automation arrived in the 1970s and 1980s, a number of industries witnessed the re-integration of tasks as machines took control of tasks associated with repetition, precision, and drudgery to raise efficiency while eliminating defects. Automated machinery took advantage of standardization, interchangeability, and flow using conveyor belts to expand further mass production. Examples include the displacement of manual workers handling die-attachment and die-bonding in semiconductor assembly with automated machinery driven by Electron Beam Induced Current (EBIC) (Rasiah, 1988). In contrast to arguments on the deskilling consequences from these changes, Rasiah (1989) provided evidence to show that direct worker skills changed from dexterity-intensive to numerical and technical skills. While employment per firm fell, productivity rose sharply in these firms. Such a structural unemployment effect created was resolved through expansion in other industries so that unemployment levels in Malaysia fell from 6.7 percent in 1985 to 4.0 percent in 1990.

As with the Luddite movement, the proliferation of IR4.0 technologies has inevitably invoked concerns among workers and worker organizations over the future of work and employment. Humanists anticipate the transformation of work further away from deskilling tendencies to one enriching further humans' mental faculties that will allow workers to realize their creative self. If Freire (1970) had used the metaphor of 'day in day out, if workers are just loading bricks onto wheelbarrows before pushing and eventually unloading them at construction sites, which is something that horses or donkeys could do', changes in work now can be expected to offer workers the opportunity to evolve their innovative capabilities to realize their creative selves. However, such a transformation brings with it serious challenges, i.e., unless the workers have trained themselves to evolve their mental faculties and governments are ready to finance their knowledge-based retraining, the flipside of new technologies, of creating mass unemployment, may become the order of the day. At its extreme end, the lack of resources could prevent poor countries, as well as poor people in middle-income and rich countries from enjoying access to such technologies thereby exacerbating the digital divide. Both experiential and tacit knowledge are critical, and though their early articulations have been philosophical with examples that are dated now, the new manifestations do not break the boundaries of such conceptualizations just as the structure and concept of the computer is still visible in Babbage's 1868 articulation of it (Britannica, 2020).

While scholars are divided in the way technical change is conceptualized (see Rasiah, 2019), they generally concur that the embedding ecosystem is critical to stimulate the evolution of firms connecting with digitalization to adopt IR4.0 practices. Consequently, governments have launched roadmaps to accelerate the development of the digital infrastructure. Malaysia is no exception here as the Digital Free Trade Zone was launched by the prime minister in 2017 with the assistance of Jack Ma (Bhunia, 2017), while the IR4.0 Master Plan (Malaysia, 2018) and the Malaysian Digital Economy Blueprint (Malaysia, 2020) were launched in 2018 and 2020, respectively (see also Malaysia, 2019). The latter two are promotional plans effectively targeting digitalization and absorption of IR4.0 technologies till 2030. Significant efforts have since been taken by particular ministries to implement digitalization initiatives with development as the focus. However, most of these efforts have not tangibly shifted from readiness assessment to an actual action plan to quicken its execution, including sequencing to ensure its effective and efficient implementation.

While one can quibble with some aspects of the Malaysian blueprints on digitalization and industry 4.0 technologies, it does show the government's seriousness to catch the wave of change through proactive policies. Both the Twelfth Malaysia Plan (2021–2025) and the 2022 budget reflect some of their proactive initiatives (Malaysia, 2021a, 2021b). The Malaysian government has invested in a number of programmes that provide access to digital technology and to also provide the necessary basic services (including for the upgrading of digital knowledge and skills). The government also recognizes the need to bridge the digital divide through meaningful policies that address issues of empowerment. As we embrace this digital era, we need to move the economy forward to achieve the goals identified in Malaysia's Shared Prosperity Vision blueprint through a focus on quickening innovations as part of a comprehensive digital economy directed at driving the country towards a prosperous, egalitarian, and sustainable nation (see Malaysia, 2019, 2020). This book covers aspects of the digital infrastructure, digital literacy and cyber security, which is timely and pertinent as a platform to deliberate the various issues related to digital transformation in the era of IR4.0.

While technological advancement through digitalization and IR4.0 technologies has extended further the unbound Prometheus, [2] it is simply a too broad concept to target its capture in a single book. Hence, inevitably this book does not seek to present an exhaustive account of the various sectors, and activities through which digitalization and IR4.0 technologies will impact on Malaysia's development. Instead, the book focuses on selected critical sectors where the ecosystem for its implementation and synergistic development need strengthening to usher a comprehensive appropriation of the associated synergies.

Critical Components

While digitalization promises the diffusion of IR4.0 technologies, nations can open up the discourse to enable the open development of such technologies, but because economic resources are scarce, many will have to focus on a few critical sectors to accelerate its emergence and expansion optimally. Consequently, we focus here on a selected range of focal areas that will become important pillars for Malaysia's development. In doing so, considerations are given to appropriating collective synergies, solving collection action problems, stimulating inter-sectoral interdependence, and structural complementarities. Hence, the book seeks to focus on manufacturing, agriculture, services, the connectivity and coordination with intermediary organizations and government, the shift from mass production to mass customization, and the importance of addressing cybercrime.

Agriculture

Farming has benefitted enormously from the introduction of robots and drones, both of which rely heavily on the digital infrastructure, including internet of things (IOT), cloud computing, fibre optics cables, and big data analytics. Information on climate and weather, distant controls, and continuous improvement requires connectivity and coordination with knowledge nodes, such as research labs in universities, and incubators and science parks; the deployment of robots and drones often require application-specific adaptations. Unmanned robots and drones have since increasingly become important in farming (Raj et al., 2021).

While large-scale farming can internalize the costs to amortize investments necessary to acquire robots and drones, small and medium farms have evolved collaborative links to appropriate economic synergies at the community or cooperative level. The government has forged such collaborative links in some locations, such as Taiwan, while in Spain, France, and Italy; some unions of farmers have evolved cooperatives among farmers and artisans.

Manufacturing

The early proliferation of digitally controlled appliances entered manufacturing, which can be traced to the 1980s and 1990s. Electronic and automobile firms began to appropriate economic synergies from subjecting production schedules to just-in-time operations that connected integrated materials requirement through the operations of cellular manufacturing computer-driven controls that translated demand – both quantitative and qualitative – to the production operations.

Indeed, starting with electronics firms first, the need for access to final markets to track and test latest technologies through user-producer relationships gradually disappeared as firms began tapping R&D and designing expertise from locations offering grants and the R&D personnel, which explains the spread of frontier wafer fabrication and R&D activities at distant sites, such as Israel, Russia, and Germany, and subsequently the emergence of contract manufacturing in such sophisticated fields in Taiwan from the 1980s (Rasiah & Yap, 2019).

The diffusion of digitalization into automobile industries followed next as firms in the industry attempted to match demand-supply coordination both quantitatively and qualitatively to eliminate surplus inventories and defects. Inter-country developments, such as the small group activities focused on kaizen among Japanese firms (e.g., Toyota, six sigma management practices) in American firms (e.g., General Motors), and works councils (e.g., Volkswagen) seeking to appropriate fair reward for innovations among German firms, and their buyers and suppliers had already driven the search for upgrading among the frontier firms from the 1970s.[3] These practices emerged strongly in semiconductor firms too.

With production largely outsourced to contractors located among the developing countries, the fortunes of the textiles and clothing industry fluctuated between the big buyers who also held the brand names (Gereffi, Humphrey & Sturgeon, 2005), and their links with second-tier contractors with original equipment manufacturing (OEMs) and original designing manufacturing (ODMs). It is this set of firms that grew in the 1960s and 1970s in Hong Kong, Singapore, South Korea, Taiwan, Malaysia, Brazil, and Mexico, which largely organized operations in their own countries, as well as other low-cost locations, initially among countries such as, Jamaica, Peru, Bangladesh, Cambodia, Lesotho, and Kenya to access quotas in developed country markets. A combination of the need to manufacture high-quality fabric and garments, and the increased efficiency and effectiveness offered by robots, and centrally controlled systems, which increasingly came from buyers led to the increasing penetration of digitalization among textiles and clothing firms. The shift to robotization in the textiles and clothing industry was quickened by new initiatives to impose trade restrictions with President Trump raising tariffs on imports in general, and China in particular from 2017. The latter drove Taiwan's manufacturing and agricultural firms to reshore operations from China to Taiwan. The COVID-19 pandemic exacerbated the situation as firms sought to align around key markets. The Japanese government announced a grant of USD 2.2 billion in 2020 to encourage Japanese firms to relocate from China to Japan and Southeast Asia (Denyer, 2020). Also, lead firms' such as Tian Yuan Garments from China and Taiwan Semiconductor Manufacturing Corporation from Taiwan relocated manufacturing using robotized technologies in Arkansas in 2018 and Phoenix in 2021, respectively.

Services

Two pillars of services are problematized here, *viz.*, digital education, and digital trade and e-commerce. Both are undergoing radical transformation in many parts of the world, and Malaysia is no exception.

Digital Education

Three major aspects of education related to digitalization and IR4.0 are emphasized in this volume. The first deals with robots as teachers, while demand deals with changes effected by transformations sought by socioeconomic agents and intermediary organizations, and the third relates to supply responses. With the first, artificial intelligence (AI)[1] has enabled automated machines to undertake complex tasks with little or no supervision. For example, autonomous robots can sense, think, and act without any external control (Lin, Abney & Bekey, 2014). Often such robots are governed by function, and when they take anthropomorphic form, they function as humanoid robots as they are then programmed to behave somewhat like people, including in the way they communicate (Newton & Newton, 2019).

In the second, demand for digitalization and IR4.0 technologies has continued to transform the way socioeconomic agents and intermediary organizations have sought to connect with the various knowledge nodes from where users (including school students) can influence changes in the creation of new modes of interactive learning and acquisition of knowledge.

The provision of education through digital modes is one major aspect of digital education. Malaysia, South Korea, and Taiwan, for example, launched smart-schools programmes in 1999, 2012, and 2014 respectively (Rasiah, 2019). YTL, Samsung, and TSMC spearheaded the introduction of those programmes in Malaysia, South Korea, and Taiwan, respectively. Smart programmes in some schools have continued to stimulate smartification by stimulating socioeconomic participants to innovate in the networks both individually and collectively while utilizing such platforms. This is the classic knowledge network in which the more the number and active participants – engaged in both using and contributing from and to the knowledge pool – the higher the synergies that can be appropriated by the participants in the network.

In addition to the laying of broadband infrastructure, and supplying computers, and internet connections, the supply of aspects of digitalization and IR4.0 into education also entails the proactive linking of producers and users in the information and knowledge networks. The proliferation of distance learning operatives has a long history, but the emergence of such a platform arising from digitalization has increasingly become important since the turn of the millennium. While problems with travel traffic jams and competing demands for economic agents saw the increased use of zoom platforms for presentations and meetings, the outbreak of the COVID-19 pandemic forced its wide use as not only the need of social distancing and handling increased travel costs, and visa restrictions raised the use of these platforms across the world. A large number of students were attending classes since the COVID-19 pandemic broke out in 2020 through digital support systems at several schools, colleges, and universities from different corners of the planet.

Digital Trade and E-commerce

The promotion of information and communication technology (ICT) in commerce emerged quickly after the digital revolution began in the 1990s. In addition to firms, government caught the waves of change early to promote e-government and other services linking, inter alia, commerce. The National Information Technology Agenda was started in the Seventh Malaysia Plan (Malaysia, 1996). Information services became a major vehicle of development in the Eighth Malaysia Plan (Malaysia, 2001). The focus on ICT rose further in the Ninth, Tenth, and Eleventh Malaysia Plans (Malaysia, 2006, 2011, 2016), while the IR4.0 national Policy and digital economy blueprint were launched in 2018 and 2020, respectively (Malaysia, 2018, 2020). E-commerce took a higher dimension when the national e-commerce roadmaps 1.0 and 2.0 were launched in 2016 and 2021, respectively.

As social distancing became a key precautionary instrument to check the spread of the COVID-19 pandemic since 2020, it escalated further digitalization as the use of e-commerce intensified to meet the disruptions associated with restrictions on movements. Indeed, trade and finance are major channels through which digitalization and digitization is transforming the landscape of business and commerce.

Expansion of Shared Economy

Digitalization and IR4.0 technologies have also transformed consumer conduct in economies as the increasing capacity to mass customize as given the opportunity for the provision of wider choice and yet at cheaper prices. Its coverage ranges from enabling its access and reach to encompass. Indeed, the advent of platform for use by robots and drones has led to a shift in focus on mass production to include mass customization so that both scale and scope economies can be appropriated at the same time.

The sharing of common platforms has stimulated the expansion of the shared economy. Consequently, there has been an upheaval in the birth and expansion of several businesses in the gig economy. Common platforms are being increasingly shared by a wide range of digital businesses to deliver services to both small producers and consumers thereby driving mass customization (Pech & Vrchota, 2022).

Intermediary Organizations

Given the public goods and public utilities properties of a number of the activities that digitalization supports, governments have an important role to play to support the development of the digital infrastructure to regulate its activities.[4] In addition to the installation of broadband infrastructure, governments are active on assisting the appropriation of economic synergies by offering support services to schools, firms, hospitals, and other users. In addition, governments also provide the enabling environment to stimulate the development of robot and drones so as to promote the use of gadgets and vehicles that merge augmented and virtual realities. Furthermore, governments have revised the filing and use of intellectual property rights (IPRs) to stimulate its diffusion through incremental innovations, and in some cases, governments have tried to hold IPRs that serve the public. In addition, governments have been instrumental in strengthening the security system to prevent cybercrime.

The next important function of the government is to prepare the labour force for the range of skills and technical knowledge that is required to participate in the IR4.0 revolution. In addition to schools, colleges, and universities, it is critical that the vocational and technical institutes catch the wave of change to expose trainees to the new skills expected of them. Given the commercial opportunities associated with the new skills a facilitate environment is often the alternative governments help create so that the private sector becomes a key engine that orientates the direction of training and learning.

While to a large extent the clustering of firms and organizations has evolved without massive government intervention at Route 128 and the Silicon Valley in the United States, such connected knowledge networks have evolved in other places such as Taiwan, South Korea, Japan, Singapore, and Taiwan through strong government support (Best, 2001, 2018). Government strengthened the large chaebols in Korea, with these conglomerates. Such a knowledge network that connects individuals, firms, organizations, (including science and technology parks that house incubators), and agencies entrusted with regulations is often referred to as the science, technology, and innovation (STI) infrastructure.

Malaysia is at a stage of development where critical technologies ought to be rooted in the country to support sustainable development, which must include a strong base to adapt and develop technologies associated with strengthening digitalization and IR4.0. While there are already-independent firms that have emerged to design and fabricate robots and drones, (such as DreamEdge and Aerodyne), and some research universities (such as University of Malaya and Universiti Teknology Malaysia) have set up IR4.0 centres to support firms seeking solutions to their digital problems, and to design and develop robots and drones for them, the development as yet needs quickening. There will need to be a mobilization of these centres and firms, and science and technology parks in Malaysia to support R&D and commercialization of digitally supported IR4.0 instruments. It is not just enough to construct and launch these organizations; there must be efforts to connect these organizations with individuals and firms, and with supporting government agencies.

The connectivity and coordination will inevitably have to create networks that link socioeconomic agents with intermediary organizations to solve collective action problems, and to appropriate socioeconomic synergies from the collaborative participation of agents and organizations in networks as a whole (Figure 1.1; see also Rasiah, 2019).

The final sub-pillar of intermediary organizations examined is cybersecurity issues arising from the mass expansion of internet and digital connections. While cybersecurity is the responsibility of all stakeholders, governments have taken the lead to spearhead the regulation of cybercrime that stretches from data theft, cyber-bullying, hacking, and robbery (Zaballos & Harranz, 2013; Grabosky, 2015; Holt, Bossler & Siegfried-Speller, 2017). The exponential expansion of

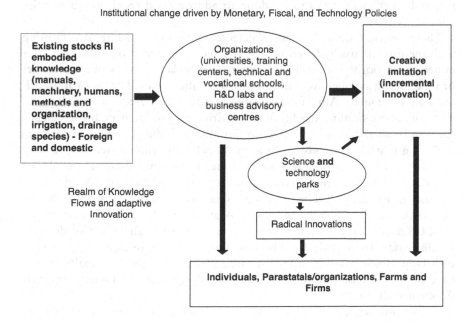

Figure 1.1 Open System of Innovations.
Source: Adapted from Rasiah (2019).

participants with complex programmes has made the tracking of cybercrime difficult. Nevertheless, it is critical that clearly laid out laws, responsible policing, and enforcement authorities are put in place to regulate the cyberspace to discourage and prevent crime. Consequently, critical issues related to cybersecurity and cybercrime in Malaysia is examined in the book.

Outline of book

Following this introduction chapter, the rest of the chapters examine the extent of institutional support that has diffused through the Malaysian economy, and what more should be done to strengthen the ecosystem to quicken the diffusion of IR4.0 technologies in the country. This is done through the broad categories of sectors (i.e., agriculture, industry with a focus on manufacturing, and services), and the intermediary organizations to drive them.

Agriculture

While the contribution of agriculture to GDP in Malaysia has declined over the decades since independence in 1957 but especially since 1970, falling productivity and chronic trade deficits especially in food production since the late 1980s has raised serious concerns. Chapters 3 and 4 focus on small-scale intensive farming, and large-scale extensive farming. Much of the discussion in these chapters examine the extent of interventions henceforth in the country, and what more needs to be done to address food security and to raise competitiveness.

There are significant differences between extensive and intensive farming with the former largely driven by internalized-scale operations, and the latter either by individualized small farm operations specializing on scope economies or functioning as collective units around a number of small farms to appropriate scale economies. Also, food farming is often associated with small farms where intensive farming has become the driver of productivity and efficiency gains in Israel and Taiwan. Cash crop farming, including commodity farming with long gestation periods, (such as in oil palm and rubber tree cultivation), is often associated with large-scale farming. Even poverty alleviation driven small scale cash crop farming, as undertaken by the Federal Land Development Authority (FELDA), Rubber Industry Smallholders Development Authority (RISDA), and the Federal Land Consolidation and Rehabilitation Authority (FELCRA) appropriate scale-efficiency synergies through collective decision making under these centralized bodies (Rasiah, 2018). In essence, large-scale cash crop farming is dominated by large plantations, though smallholdings have evolved. Consequently, each chapter is devoted to the two significantly different-scale operations.

On the one hand, in Chapter 2, Afzanizam first examines how digitalization and IR4.0 technologies can raise food production through their deployment in fertigation. Such intensive farming methods not only help synchronize the flow of fertilizers, water, and other materials in food farming that not only raise efficiency and remove wastage but its effective coordination makes small-scale farming meet its objective of raising yields (Afzanizam, 2023). In Chapter 3, Suresh Palpanabhan examines the proliferation of IR4.0 technologies in large-scale farming by looking at national policies that promote such technologies, as well as its extension to enable its use in other productivity-driven large-scale farming activities with a focus on both robots and drones. In addition to a critical review of existing policies, the chapter also recommends the expansion of IR4.0 activities in large-scale farming (Suresh, 2023).

Manufacturing

This sub-section examines the embedding environment facing two major export-oriented manufacturing industries in Malaysia, and if digitalization helped mitigate the spread of COVID-19 in the manufacturing sector. Chapter 4 by Yeap Khai Leang and Rajah Rasiah examines the presence and quality of the embedding environment supporting the diffusion of IR4.0 in electronics (Yeap & Rasiah, 2023). The evidence that they produce shows significant initiatives to promote the adoption of IR4.0 technologies but argue that the quality of connectivity and coordination between the supporting organizations and electronics firms need major improvements.

Chapter 5 by Kiranjeet Kaur, Hemant Kedia, and Rajah Rasiah examines the state of the ecosystem supporting the utilization of IR4.0 technologies in the textiles and apparel industry in Malaysia (Kiranjeet, Hemant & Rasiah, 2023). Often mis-classified as a sunset industry, this chapter discusses significant developments that firms in the industry have embarked on to absorb IR4.0 technologies, but their quests would require the ecosystem system embedding them to be strengthened for them to appropriate further economic synergies from such technologies.

Taking on the suggestions by international organizations, such as the World Bank, the United Nations Conference for Trade and Development (UNCTAD), and the International Monetary Fund (IMF), Shankaran Nambiar and Yip Tien Ming assess in Chapter 6 the extent to which digitalization has helped mitigate the spread of COVID-19 cases in the manufacturing sector. The econometric methodology they deploy leads them to conclude that digitalization has helped reduce the negative effects of COVID-19 by 98 percent in Malaysia's manufacturing sector (Nambiar & Yip, 2023).

Services

The services sector is highly complex with several sub-sectors, which are so wide and diverse that it is difficult to include all of them. Consequently, this book focuses on three critical areas, i.e., education, e-commerce, and the gig economy. In doing so, avoided discussion on such critical areas such as power, water, and healthcare. Chapter 8 by Nurliana Kamaruddin, Cecilia Cheong Yin Mei, Wah

Yun Low, and Azirah Hashim examines the critical issues and challenges facing digital literacy in Malaysian education. Taking on the argument that traditional literacy is inadequate address the literacy demands imposed by digitalization, the chapter discussed a number of motivating factors for educators and learners to make the migration from traditional to online teaching and learning possible. In doing so, the chapter emphasizes the problems posed by incoherent policy-making at the national and institute levels, the digital divide that exists, the readiness of teachers and students in making the transition, the concerns facing special needs education and, security concerns digitalization poses (Kamaruddin, Cheong, Low & Azirah, 2023).

Chapter 7 by Dorothy DeWitts and Norlidah Alia examines creative digital pedagogies for student engagement and preparing Malaysian students for the future. The chapter argues that Malaysian students need to be equipped with digital skills with a focus on creative and innovative thinking to embrace smoothly economic transformation to digitalization and digitization. In doing so, that chapter emphasizes on digital creativity as a critical for the Malaysia's future workforce, and to prepare teachers and instructors the pedagogies and skills required to encourage creative behaviour among students. The chapter recommends improving the technology pedagogical content knowledge among teachers and instructors based on an integrated digital competency framework, which include empowering learners, creating new knowledge, strengthening connectivity in communities, and designing solutions and computational thinking. The chapter also recommends the use of collaborative applications for learning and virtual reality, as well as problem-solving approaches using visual programing applications (DeWitts & Norlidah, 2023).

Chapter 9 by Tham Siew Yean and Andrew Kam Jia Yi examines the promotion of e-commerce through the expansion in ICT and digitalization. The chapter specially examines the nature of e-commerce in Malaysia, including cross-border e-commerce and identifies the key challenges the country is facing in moving forward (Tham & Kam, 2023). Khairul Hanim Pazim, Yosuke Uchiyama, Fumitaka Furuoka, Beatrice Lim, and Jingyi Li examine in Chapter 10 the emergence and expansion of the shared economy in Malaysia following the proliferation of digitalization and the gig economy. The chapter discusses the manner with which the gig economy is unfolding, which is driven by digitalization but one that requires economic agents to share common platforms (Khairul Hanim et al., 2023).

Intermediary Organizations

Given the critical role intermediary organizations play in ecosystems that support individuals, farms, and firms' activities, four chapters in this book focus on the state of support socioeconomic agents enjoy from such organizations. Chapter 13 by Abdul Latif and Saliza Saari discusses governance instruments introduced by a key standards and rules organization, i.e., the Malaysian Productivity Corporation (MPC) targeted at stimulating productivity improvements of economic agents in Malaysia (Abdul Latif & Saliza, 2023). In Chapter 11, Cheong Kee Cheok and Li Ran address critically the landscape facing technical and vocational education training (TVET) that is pertinent to produce the digitally competent workforce to support the expansion in IR4.0 technologies in the Malaysian economy. The chapter makes it a point to flag the need to align policies with the education and training organizations that are in sync with the emerging skills in supporting IR4.0 technologies (Cheong & Li, 2023).

In Chapter 12, Naguib Mohd Nor and Nazreen Mohd Nasir articulate the role played by STRAND in stimulating the absorption of IR4.0 technologies in Malaysian Manufacturing Industries. STRAND is a special organization appointed to address market failures and to coordinate public sector initiatives to interface smoothly with private sector efforts at introducing IR4.0 technologies (Naguib & Nazreen, 2023). Chapter 14 by Maslina Daud and Rajah Rasiah discusses cybersecurity breaches and efforts taken by the government and other bodies to prevent cybercrime. This is an important chapter that emphasizes the significance of security instruments to ensure the smooth functioning of the digital system that is evolving to, among other things, support the operation of AI instruments (Maslina & Rasiah, 2023).

Overall, the chapters in the book capture the technocratic dynamics of change driven and the concerns digitalization and IR4.0 have raised through a balanced assessment of its impact on the Malaysian economy by focusing on the embedding ecosystem. It is a rare contribution that examines its proliferation in the major sectors of agriculture, manufacturing, and services, as well as the intermediary organizations essential for the promotion and protection of the processes and orderly performance of the embedding infrastructure and socioeconomic agents. The book breaks ground by reviewing incisively policy instruments critical for quickening the effective and orderly development of the embedding organizations and regulatory framework to quicken the appropriation of socioeconomic synergies from digitalization and industry 4.0 technologies to ease the catching of the gales of creative destruction in general, and Malaysia in particular.

Notes

1 One has to distinguish this from flexible casualization in which flexiblization leads to a rise of the informal sector that is exposed to little or no training and low wages (Rasiah, 2009; Ofreneo, 2009).
2 See Landes (2003) for his account of the unbound Prometheus from 1750 till the turn of the millennium.
3 Wormack et al. (1990) only capture firm-level developments that was already evolving strong by the 1980s (Rasiah, 1988, 1995).
4 Public goods are non-excludable (the inclusion of additional socioeconomic agents cannot be precluded) and non-rivalrous (once created no additional costs involved). Knowledge, security, and the environment are examples of the public goods. However, public utilities are both excludable and rivalrous, but they need to reach most members of society. Examples are water and electric supply, and healthcare.

References

Abdul Latif & Saliza, S. (2023). Government Initiatives to Promote Adoption of IR4.0 Technologies in Manufacturing, Rasiah, R., Low, W.Y. & Kamaruddin, N. (eds) *Digitalization and Development: Ecosystem for Promoting Industry 4.0 Technologies*, (cross referenced from same book).

Afzanizam, M.A.R. (2023). Digitalizing Food Production through the Fertigation System, Rasiah, R., Low, W.Y. & Kamaruddin, N. (eds) *Digitalization and Development: Ecosystem for promoting Industry 4.0 Technologies*, (cross referenced from same book).

Andrews, E. (2019). Who were the Luddites, History Stories, June 26 updated from the original published on August 7, 2015, Downloaded on August 24 from https://www.history.com/news/who-were-the-luddites.

Arai, Y., Naganuma, S. & Satake, Y. (2012). Broadband deployment projects in less-favored areas and the broadband policies of national and local governments in Japan, *Komaba Studies in Human Geography* 20, 14–38. (In Japanese).

Arai, Y. (2019). History of the Development of Telecommunications Infrastructure in Japan, *Networks and Communications Studies*, 33(3), 4.

Best, M.H. (1990). *The New Competition*, Cambridge: Harvard University Press.

Best, M.H. (2001). *The New Competitive Advantage*, Oxford: Oxford University Press.

Best, M.H. (2018). *How Growth Really Happens: The Making of Economic Miracles through Production, Governance, and Skills*, Princeton: Princeton University Press.

Bhunia, P. (2017). Digital Free Trade Zone Launched in Malaysia by Najib Razak and Jack Ma, OpenGov, October 28, accessed on August 26 from https://opengovasia.com/digital-free-trade-zone-launched-in-malaysia-by-pm-najib-razak-and-jack-ma/

Britannica, (2020). The Editors of Encyclopaedia. "Charles Babbage". *Encyclopedia Britannica*, 28 December, accessed on August 26, from https://www.britannica.com/biography/Charles-Babbage.

Cheong, K.C. & Li, R. (2023). The Role of TVET in IR4.0 for Malaysia, Rasiah, R., Low, W.Y. & Kamaruddin, N. (eds) *Digitalization and Development: Ecosystem for Promoting Industry 4.0 Technologies*, (cross referenced from same book).

Denyer, S. (2020). Japan helps 87 Companies to Break from China After Pandemic Exposed Overreliance, *Washington Post*, July 21, accessed on June 28, 2022 from https://www.washingtonpost.com/world/asia_pacific/japan-helps-87-companies-to-exit-china-after-pandemic-exposed-overreliance/2020/07/21/4889abd2-cb2f-11ea-99b0-8426e26d203b_story.html.

DeWitts, D. & Norlidah, A. (2023). Creative Digital Pedagogies for Student Engagement: Preparing students for Industry 4.0, Rasiah, R., Low, W.Y. & Kamaruddin, N. (eds) *Digitalization and Development: Ecosystem for Promoting Industry 4.0 Technologies*, (cross referenced from same book).

Freire, P. (1970). *Pedagogy of the Oppressed*, New York: Continuum International Press.

Gereffi, G., Humphrey, J. & Sturgeon, T. (2005). Governance of Global Value Chains, *Review of International Political Economy*, 12(1), 78–104.

Grabosky, P. (2015). *Cybercrime*, Oxford: Oxford University Press.

Holt, T.J., Bossler, A.M. & Siegfried-Speller, K.C. (2017). *Cybercrime and Digital Forensics: An Introduction*, London: Routledge.

Huawei (2021). *Intelligent World 2030*, Shenzen: Huawei.

Kamaruddin, N., Cheong, C.M., Low, W.Y. & Azirah, H. (2023). Rasiah, R., Low, W.Y. & Kamaruddin, N. (eds) *Digitalization and Development: Ecosystem for Promoting Industry 4.0 Technologies*, (cross referenced from same book).

Kiranjeet, K.S., Hemant, K. & Rasiah, R. (2023). Ecosystem for Promoting IR 4.0 Technologies in Textiles and Clothing Manufacturing, Rasiah, R., Low, W.Y. & Kamaruddin, N. (eds) *Digitalization and Development: Ecosystem for Promoting Industry 4.0 Technologies*, (cross referenced from same book).

Landes, D.S. (2003). *The Unbound Prometheus: Technological Change and Industrial Change from1750 till the Present*, Cambridge: Harvard University Press.

Lin, P., Abney, K., & Bekey, G.A. (2014). *Robot Ethics: The Ethical and Social Implications of Robotics*, Cambridge: The MIT Press.

Malaysia (1996). *The Seventh Malaysia Plan 1996–2000*, Kuala Lumpur: Government Printers.

Malaysia (2001). *The Eighth Malaysia Plan 2001–2005*, Putrajaya: Government Printers.

Malaysia (2006). *The Ninth Malaysia Plan 2006–2010*, Putrajaya: Government Printers.

Malaysia (2011). *The Tenth Malaysia Plan 2011–2005*, Putrajaya: Government Printers.

Malaysia (2016). *The Eleventh Malaysia Plan 2016–2020*, Putrajaya: Government Printers.

Malaysia (2018). *National Policy on Industry 4.0*, Ministry of International Trade and Industry, accessed on August 26 from https://www.miti.gov.my/miti/resources/National%20Policy%20on%20Industry%204.0/Industry4WRD_Final.pdf.

Malaysia (2019). *Shared Prosperity Vision 2030*, Putrajaya: Malaysia National Printers.

Malaysia (2020). *Malaysia Digital Economy Blueprint*, Economic Planning Unit (EPU), accessed on August 26 from https://www.epu.gov.my/sites/default/files/2021-02/malaysia-digital-economy-blueprint.pdf.

Malaysia (2021a) *The Twelfth Malaysia Plan 2021–2025*, Putrajaya: Malaysian National Printers.

Malaysia (2021b) Malaysia's Budget 2022: Key Takeaways for Employers and HR to Note, November 1, accessed on June 28, 2022 from https://www.humanresourcesonline.net/malaysia-s-budget-2022-key-takeaways-for-employers-and-hr-to-note.

Maslina, D. & Rasiah, R. (2023). Addressing Cybersecurity Issues, Rasiah, R., Low, W.Y. & Kamaruddin, N. (eds) *Digitalization and Development: Ecosystem for Promoting Industry 4.0 Technologies*, (cross referenced from same book).

Naguib, M.N. & Nazreen, M.N. (2023). Addressing Cybersecurity Issues, Rasiah, R., Low, W.Y. & Kamaruddin, N. (eds) *Digitalization and Development: Ecosystem for Promoting Industry 4.0 Technologies*, (cross referenced from same book).

Nambiar, S. & Yip, T.M. (2023). Did Digitalization Help Manufacturers Cope with COVID19 Pandemic? Rasiah, R., Low, W.Y. & Kamaruddin, N. (eds) *Digitalization and Development: Ecosystem for Promoting Industry 4.0 Technologies*, (cross referenced from same book).

Nazia, N. & Rasiah, R. (2022). *Industrial Upgrading in the Textiles and Clothing Industry: Pakistan and Late Industrialization*, forthcoming.

Newton, D.P. & Newton, L.D. (2019). Humanoid Robots as Teachers and a Proposed Code of Practice, *Frontiers in Education*, 4(125), 1–10.

Ofreneo, R. (2009). Development Choices for Philippine Textiles and Garments in the Post-MFA Era, *Journal of Contemporary Asia*, 39(4), 543–561.

Khairul Hanim, P., Yusoke, I., Fumitaka, F., Lim, B. & Li, J. (2023). Digitalization and the GIG Economy: Ecosystem System Supporting Consumer Service, Rasiah, R., Low, W.Y. & Kamaruddin, N. (eds) *Digitalization and Development: Ecosystem for Promoting Industry 4.0 Technologies*, (cross referenced from same book).

Pech, M. & Vrchota, J. (2022). Product Customization Process in Relation to Industry 4.0 and Digitalization, *Processes*, 10(3), 539.

Raj, A.Y., Venkataraman, A., Vinodh, A. & Kumar, H. (2021). Autonomous Drone for Smart Monitoring of an Agricultural Field, *7th International Engineering Conference "Research & Innovation amid Global Pandemic" (IEC)*, 2021, pp. 211–212. doi: 10.1109/IEC52205.2021.9476097.

Rasiah, R. (1988). The Semiconductor Industry in Penang: Implications for the New International Labour Theories, *Journal of Contemporary Asia*, 18(1): 24–46.

Rasiah, R. (1989). Competition and Restructuring in the Semiconductor Industry: Implications for Technology Transfer and Absorption in Penang, *Southeast Asian Journal of Social Sciences*, 17(2), 41–57.

Rasiah, R. (1995). *Foreign Capital and Industrialization in Malaysia*, Basingstoke: Macmillan.

Rasiah, R. (2004). *Foreign Firms, Technological Capabilities, and Economic Performance: Evidence from Africa, Asia and Latin America*, Cheltenham: Edward Elgar.

Rasiah, R. (2007). The Systemic Quad: Technological Capabilities and Economic Performance of Computer and Component Firms in Penang and Johor, Malaysia, *International Journal of Technological Learning, Innovation and Development*, 1(2), 179–203.

Rasiah, R. (2009). Malaysia's Textile and Garment Firms at the Crossroad, *Journal of Contemporary Asia*, 39(4), 530–542.

Rasiah, R. (2018). *Developmental States: Land Schemes, Parastatals and Poverty Alleviation in Malaysia*, Bangi: Universiti Kebangsaan Malaysia Press.

Rasiah, R. (2019). Building Networks to Harness Innovation Synergies: Towards an Open Systems Approach to Sustainable Development, *Journal of Open Innovation: Technology, Markets and Complexity*, 5, 70

Rasiah, R. (2020). Industrial Policy and Industrialization in Southeast Asia, Arkebe, O., Cramer, C., Chang, H.J. & Wright, R.K. (eds), *Oxford Handbook of Industrial Policy*, Oxford: Oxford University Press.

Rasiah, R. & Yap, X.S. (2019). How Much of Raymond Vernon's Product Cycle Thesis Is Still Relevant Today: Evidence from the Integrated Circuits Industry, *International Journal of Technological Learning, Innovation and Development*, 11(1), 56–77.

Suresh, P. (2023). Proliferation of IR4.0 Technologies in Large Scale Agriculture, Rasiah, R., Low, W.Y. & Kamaruddin, N. (eds) *Digitalization and Development: Ecosystem for Promoting Industry 4.0 Technologies*, (cross referenced from same book).

Taylor, F.W. (1911). *Principles of Scientific Management*, New York: Harper & Brothers.

Tham, S.Y. & Kam, A.J.Y. (2023). Ecommerce Expansion in Malaysia Rasiah, R., Low, W.Y. & Kamaruddin, N. (eds) *Digitalization and Development: Ecosystem for Promoting Industry 4.0 Technologies*, (cross referenced from same book).

Tolliday, S. & Zeitlin, J. (1987). *The Automobile Industry and its Workers: Between Fordism and Flexibility*. New York: St. Martin's Press.

United Nations (2015). Sustainable Development Goals, accessed on June 28, from https://developers.google.com/community/gdsc-solution-challenge/UN-goals.

United Nations (2021). *UNESCO Science Report: The Race Against Time for Smarter Development*, Paris: UNESCO.

Womack, J.P., Jones, D.T. & Roos, D. (1990) *The Machine that Changed the World*, New York: Free Press.

Yeap, K.L. & Rasiah, R. (2023). Diffusion of IR 4.0 Technologies in Electronics Manufacturing, Rasiah, R., Low, W.Y. & Kamaruddin, N. (eds) *Digitalization and Development: Ecosystem for Promoting Industry 4.0 Technologies*, (cross referenced from same book).

Zaballos, A.G. & Harranz, F.G. (2013). From Cybersecurity to Cybercrime: A Framework for Analysis and Implementation, Technical Note, New York: Inter-American Development Bank.

2 Improving Food Production through Digitalizing the Fertigation System

Mohd Afzanizam Abdul Rashid

Introduction

The spread of COVID-19, officially categorized as a pandemic by the World Health Organization (WHO) on 11 March 2020, has wreaked havoc on the human population. The global economy was negatively impacted as the world's GDP fell by 3.2 percent in 2020 with the abrupt displacement of the labour force happening as various economic sectors were not allowed to fully operate. The Malaysian economy was also affected when its total output took a plunge in the second quarter of 2020 and the GDP declined significantly by 17.2 percent year on year. The unemployment rate climbed to 5.3 percent in May that year, bringing it to the highest recorded unemployment rate since the onset of the Asian Financial Crisis in 1997–1998.

Despite the calamities encountered, the agricultural sector and food production activities, in particular, were allowed to operate in Malaysia during the Movement Control Order (MCO), which began on 18 March 2020. The exemption is premised on the fact that food production and its related activities are essential economic activities. To this end, the agricultural sector is given due recognition for its importance to society. Arguably the pandemic has been a blessing in disguise for the sector and those who lost their jobs were able to venture into the agricultural sector to pursue new employment and business opportunities.

That said, there are clear gaps in the agricultural sector and industries that need to be addressed. One major issue is how the sector is generally associated with the plantation sector as this sub-sector makes up most of the land usage. The palm oil sector dominates any discourse on agriculture in Malaysia as there are large corporations and listed entities driving the industry (see Suresh, 2023). The investors' community are also more familiar with issues surrounding oil palm crop yield and volatility as well as crude palm oil (CPO) prices as these issues are well covered by equity analysts in the investment banks fraternity. Another issue is food security whereby there are a handful of crops and livestock produced that have a lower than 100 percent self-sufficiency ratio (SSR). Additionally, the trade deficits in food stuff have been persistently widening, rising to more than RM20 billion in 2020 from merely RM1.1 billion in 1990. In this

DOI: 10.4324/9781003367093-2

respect, the country's food security ought to be looked at in order to ensure that there can be enough food produced for the country.

While there are clear and persistent problems in the sector are crystal clear, the current and most pressing challenge is to revitalize and rejuvenate the agricultural industries. This can be done by taking into account the emergence of digitalization within the industry. Digitalization practices are critical as this will determine the efficiency and productivity of the sector given the healthy demand from the domestic market. It is especially true in the context of the supply chain as digitalization can be implemented in a comprehensive manner i.e., from the upstream to the downstream industries.

There are also new and exciting emerging trends, such as indoor vertical farming that offer a promising prospect given the limited availability of space in urban areas. It can make use of non-traditional spaces for example the abundance of office and retail space that can be converted for this purpose. All other potential innovations aside, this chapter will focus on the digitalization of the fertigation system, which has received a good response from the public. According to government agencies that were engaged in the research, the state of the fertigation system within the Malaysian agricultural sector is still very traditional in nature and digitalization is almost non-existent. The only digitalization involved is timers for intervals of fertilization and irrigation to occur during a day. Overall, there is a valid need to accelerate the development of digital innovation in agriculture in order to make it a viable and attractive business sector. A good combination of meeting basic needs and leveraging the latest technology in order to gain higher productivity would compel more Malaysians to partake in the agricultural industry, especially among the young graduates.

This chapter begins by looking at the economics of food in Malaysia with a focus on intensive farming. It traces, first, the trends in food consumption patterns vis-à-vis income levels. Second, it provides an overview of food prices in Malaysia and the correlation between food prices and inflation. Following that, the chapter explores Malaysia's lack of self-sufficiency in food production. Finally, it discusses an important step towards the adoption of digitalization in the agricultural sector. While digitalization has diffused considerably in Malaysian agriculture, this chapter looks specifically at digitalizing fertigation systems to increase and cheapen food production, including on how the interactions within the supply chain of the food industries can be further improved.

Economics of Food Production

The consumption of foods and beverages is essential for human beings in order to fulfil their energy requirement for survival purposes. Hunger and malnutrition weaken the ability of a person to learn and work. Worse, a large-scale lack of food supplies can lead to fatal consequences such as famine. Food consumption and the capability to perform any physical task are obviously inseparable. Additionally, one also needs to be mindful of having a balanced diet in order to live a healthy life. Therefore, the variety of food intake necessary for a healthy society could also broaden the demand for the different types of food that a person would wish to consume.

Food Consumption Issues

Food is a special giffen good that follows demand patterns, which is not normal. Several food products (e.g., rice) share the properties of inferior goods, i.e., their demand does not change much with reasonable changes in prices (Rasiah, 2018, Rasiah & Salih, 2019). Table 2.1 shows the consumption of food and non-alcoholic beverages in comparison to income level in Malaysia. Generally, the higher the income, the higher the amount of money allocated to consume food but the rise in food demand will fall as income rises. For example, those who earn a monthly income between RM1,999 and below would generally spend a total of RM403 on food and non-alcoholic beverages. Meanwhile, those whose monthly income is RM15,000 and above generally allocate about RM1,318 on food and non-alcoholic beverages. Table 2.2 indicates the percentage of spending that was allocated to food and non-alcoholic beverages, i.e., the lower the income, groups tend to allocate a larger percentage of their financial expenditure on food with a ratio of 28.9 percent for those whose monthly income is RM1,999 and below. In contrast, for a person who earns a monthly income of RM15,000 and above, the percentage he or she allots only accounts for about 11.3 percent of their total spending.

From this, we can observe the disparity in food spending trend between those who are in the higher income bracket versus those in the lower-income group. Obviously, food and non-alcoholic beverage products are all closely linked to the agricultural sector. In Malaysia, these include but are not limited to rice, meat, vegetables, as well as oils and fats. Therefore, there is a direct link between the consumption pattern of food products with the agricultural sector.

Based on per capita consumption (PCC), rice continues to command the largest ratio with 76.5 kilograms per year (kg/year) in 2018 followed by poultry meat at 50.9 kg/year in 2019. Within the fruits and vegetable category, coconut garnered a PCC of 22.7 kg/year while round cabbage at 5.3 kg/year during 2019. As for fisheries-related food, mackerel and shrimp were the most popular with PCC of 5.7 and 4.1 kg/year, respectively, in 2019. On a side note, it is worth noting that based on the PCC, it could be deduced that the eating habits among average Malaysian may not be entirely balanced. According to the 2019 National Health and Morbidity Survey, 50.1 percent of the Malaysian adult population were reported to be overweight (30.4%) or obese (19.7%). Therefore, there is a need to create further awareness that a healthy diet should be encouraged in order to reduce the risks of non-communicable diseases (NCD) such as type 2 diabetes, cardiovascular disease, as well as several types of cancers such as breast, large intestine, pancreas and kidney cancers.

Another important factor when looking at the economics of food is price. As with other goods and services, one must pay for the price of food in order to procure them. For this purpose, it is useful to look at the Consumer Price

Table 2.1 Consumption of Food and Non-alcoholic Beverages Based on Income Band, Malaysia, 2019

	RM1,999 & Below	RM2,000– RM2,999	RM3,000– RM3,999	RM4,000– RM4,999	RM5,000– RM5,999	RM6,000– RM6,999	RM7,000– RM7,999	RM8,000– RM8,999	RM9,000– RM9,999	RM10,000– RM14,000	RM15,000 & Above
Food & non-alcoholic beverages	403	538	646	706	750	786	836	849	884	967	1,318
Rice	32	35	37	38	41	41	42	45	45	48	62
Bread & other cereals	52	68	83	91	100	106	112	118	125	143	201
Meat	49	73	89	97	105	111	119	116	127	135	189
Fish & seafood	86	115	142	157	165	171	183	178	186	203	279
Milk, cheese & eggs	32	42	50	55	60	65	72	75	77	83	111
Oils & fats	12	17	20	22	23	25	26	27	28	29	37
Fruits	23	32	41	46	48	52	56	57	59	67	102
Vegetables	53	68	79	85	86	87	90	89	92	95	117
Sugar, jam, honey, chocolate & confectionary	13	15	19	21	23	25	28	29	31	34	46
Food products	27	39	47	52	54	55	58	60	60	68	92
Coffee, tea, cocoa & non-alcoholic beverages	25	33	39	43	45	49	50	53	54	60	83
Total monthly spending	1,395	2,066	2,690	3,205	3,695	4,124	4,613	5,083	5,457	6,666	11,709

Source: Compiled from DOSM (various issues).

Table 2.2 Consumption of Food and Non-alcoholic Beverages in Total Monthly Spending, Malaysia, 2019

	RM1,999 & Below	RM2,000– RM2,999	RM3,000– RM3,999	RM4,000– RM4,999	RM5,000– RM5,999	RM6,000– RM6,999	RM7,000– RM7,999	RM8,000– RM8,999	RM9,000– RM9,999	RM10,000– RM14,999	RM15,000 & Above
Food & non-alcoholic beverages	28.9%	26.1%	24.0%	22.0%	20.3%	19.0%	18.1%	16.7%	16.2%	14.5%	11.3%
Rice	2.3%	1.7%	1.4%	1.2%	1.1%	1.0%	0.9%	0.9%	0.8%	0.7%	0.5%
Bread & other cereals	3.7%	3.3%	3.1%	2.8%	2.7%	2.6%	2.4%	2.3%	2.3%	2.1%	1.7%
Meat	3.5%	3.5%	3.3%	3.0%	2.8%	2.7%	2.6%	2.3%	2.3%	2.0%	1.6%
Fish & seafood	6.1%	5.6%	5.3%	4.9%	4.5%	4.1%	4.0%	3.5%	3.4%	3.0%	2.4%
Milk, cheese & eggs	2.3%	2.0%	1.9%	1.7%	1.6%	1.6%	1.6%	1.5%	1.4%	1.2%	0.9%
Oils & fats	0.9%	0.8%	0.7%	0.7%	0.6%	0.6%	0.6%	0.5%	0.5%	0.4%	0.3%
Fruits	1.6%	1.6%	1.5%	1.4%	1.3%	1.3%	1.2%	1.1%	1.1%	1.0%	0.9%
Vegetables	3.8%	3.3%	2.9%	2.6%	2.3%	2.1%	1.9%	1.8%	1.7%	1.4%	1.0%
Sugar, jam, honey, chocolate & confectionary	0.9%	0.7%	0.7%	0.7%	0.6%	0.6%	0.6%	0.6%	0.6%	0.5%	0.4%
Food products	1.9%	1.9%	1.7%	1.6%	1.5%	1.3%	1.3%	1.2%	1.1%	1.0%	0.8%
Coffee, tea, cocoa & non-alcoholic beverages	1.8%	1.6%	1.4%	1.3%	1.2%	1.2%	1.1%	1.0%	1.0%	0.9%	0.7%

Source: Estimated from data compiled from DOSM (various issues).

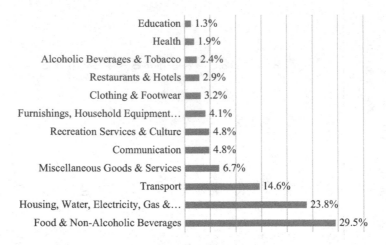

Figure 2.1 Composition of Consumer Price Index, Malaysia, 2020.
Sources: DOSM (various issues); CEIC Database (2022).

Index (CPI) which is the common measurement of the general price level in the macroeconomic setting. Judging from the CPI weighting, there is heavy importance given to food and non-alcoholic beverages with a total weightage of 29.5 percent (Figure 2.1). It is the largest share in the CPI computation which suggests that any change in the food price level will have an immediate impact on inflation.

Growing Dependence on Imports

Malaysia has been consistently dependent on foreign sources for its domestic food requirement. This may seem absurd considering the vast amount of arable land that the country has and the weather condition that are suitable for many different types of crops. The problem in Malaysia is that the agricultural sector has been skewed towards cash crops such as palm oil whereby these trees can remain productive for a period up to 25 years after 3 years of planting. Palm oil arguably has better commercial value as opposed to food crops which typically have a shorter lifespan sometimes amounting to less than a year. Therefore, replanting exercise for food crops can be daunting, especially when dealing with higher prices of fertilizers and insecticides.

Not to mention the risk that farmers may experience from unpredictable weather conditions such as heavy rains and floods as well as the heavy wind which can destroy the crops. Currently, plantation land accounts for 5.8 million hectares of Malaysia's agricultural land cultivation as opposed to food crops that command only 1 million hectares. Given that the agro-food industry does not have a comparative advantage, the quick solution would be to resort to importing.

Data on international trade for agro-food shows how much the country is dependent on the foreign food sources. The total exports and imports for foodstuff stood at RM33.8 billion and RM55.5 billion, respectively, in 2020 (Figure 2.2). This gave Malaysia a trade deficit of RM21.7 billion in that year. Comparatively, in 1990, the total exports and imports were at RM3.5 billion and RM4.6 billion, respectively, and the corresponding trade deficit was merely RM1.1 billion. In ten years, the country's trade deficit has increased by an astonishing 1821 percent. Foodstuffs that were severely in deficit are cereals and cereals preparations (RM6.6 billion); vegetables and fruits (RM6.5 billion); meat and meat preparations (RM3.3 billion); sugar, sugar preparations and honey (RM2.7 billion); dairy products and birds eggs (RM2.6 billion); feeding stuff for animals excluding unmilled cereal and fish (RM1.9 billion), as well as crustaceans and molluscs and preparation thereof (RM1.2 billion).

The SSR gives a clearer picture of the county's predicament in relation to the agro-food sector. For instance, the SSR for rice stood at 69 percent as of 2019, which is well below the 11th Malaysian Plan (11MP) target of 100 percent by 2020 (Malaysia, 2011, 2016. Similarly, the production of beef which was targeted to have an SSR of 50 percent by 2020), only managed to record 22.2 percent in 2020. Clearly, the target sets were off by a wide margin, as a result of weak implementation plans. Meanwhile, the SSR for round cabbage, chilli and ginger have been recording a declining trend from 61.9 percent, 49.1 percent and 24.0 percent in 2016, respectively, to 37.5 percent, 30.9 percent and 18.9 percent in 2020 (Figure 2.3).

This also suggests that the production of certain agro-food products has not only been unable to achieve the stipulated targets but it has also suffered from a lack of supplies, which has exacerbated the need to rely on imports. Arguably, this may have compounded the rise in food prices, especially at a time when the Malaysian ringgit has been demonstrating a weakening trend in its exchange rate against the US dollar. Therefore, increasing the number of food supplies by way of higher production levels could help address issues relating to rising food prices.

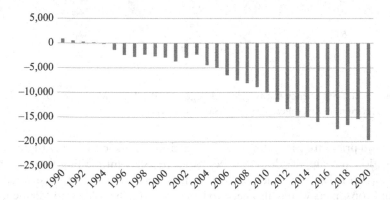

Figure 2.2 Trade Balance in Food Products, Malaysia, 1990–2020 (RM Million).
Sources: Plotted from DOSM (various issues); CEIC Database (2022).

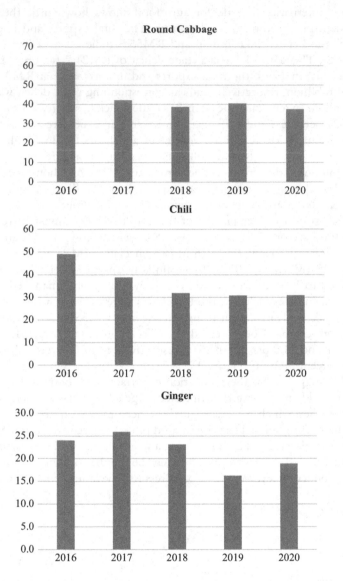

Figure 2.3 Self-Sufficiency Ratio, Selected Vegetable Products, Malaysia, 2016–2020 (%).
Source: Plotted from data collected from DOSM (various issues)

As food prices are sticky downward, it needs to be made clear that higher food production and supplies does not necessarily translate into falling food prices in the short run. The intricacies of the agro-food industries from the farm-gate to the end consumers would also need to be looked at. The role of middlemen, the market structure, and malpractice among the businesses such as price manipulation, hoarding and smuggling could also compromise the price discovery mechanism for food items. Consequently, increasing food production would be the first step to addressing the rising food prices.

The Fertigation Production Method

Given the high dependency on import for food, there have been efforts to improve local food production, especially in areas relating to the production of vegetables and fruits. While increasing production appears to be a quick fix, ensuring higher crop yields is also equally important. In this regard, there are multiple agricultural techniques when it comes to crop-growing that combines higher production growth and better productivity. One of the most common methods is fertigation which is the combination of fertilizers and irrigation. Such a technique has existed since the 1960s when drip irrigation was developed in Israel due to the shortage of water for desert agriculture. Fertigation does not require a huge landmass as crops will be planted inside polybags. Essentially, it is a soilless system whereby substrates and media such as rockwool, perlite, vermiculite, or peat are used in a polybag.

The elimination of soil improves the yield by preventing soil-borne diseases. It also increases multiple growing cycles without the need to replenish nutrients and soil conditioning. Crops such as chilli, rockmelon, and brinjals are some of the typical agricultural products that would use such a technique. Fertigation is also convenient as both fertilizers and irrigation will occur at the same time at regular intervals. This, in turn, would result in minimal use of labour as the fertilizers and irrigation will be done via electrically powered water pumps. Both fertilizers and water will go directly to each polybag, resulting in an effective way of ensuring that each plant would receive a consistent amount of nutrients at regular intervals (Figure 2.4). The fertilizers and water will follow through the irrigation pipe which will finally reach the drippers that will be placed in each polybag. This way, the amount of fertilizer and water flow is targeted for each plant, giving them sufficient nutrients that will help improve the crop yield.

Clean water can be sourced from water pipes although some may argue that using river water can be more cost-effective. However, one would need to ensure that such a water source would have the right PH level. Therefore, using a water pipe would be more convenient as it will ensure stability and reliability in terms of its supply. The users would need to set the frequency using a timer whereby regular intervals for fertilizers and irrigations would be fixed at a certain timeframe. Normally, it would take 2–3 minutes for each session to complete with a total of five sessions in a day. Before starting the process, the user would need to ensure that there is sufficient fertilizer and water in the water tank.

The AB mixed is the most common fertilizer used in the fertigation technique. It contains two parts made up of A which consists of Calcium Nitrate, Potassium Nitrate, and Fe-EDTA while B comprises Potassium Dihydrofosphate, Ammonium Sulphate, Potassium Sulphate, Magnesium Sulphate, Cupri Sulphate, Zinc Sulphate, Boric Acid, Manganese Sulphate and Ammonium Hepta Molybdate. Before mixing, fertilizer A and fertilizer B are dissolved into water separately.

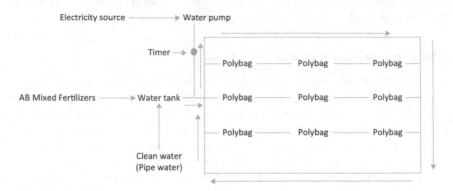

Figure 2.4 Fertigation Flow Chart.
Source: Author.

Otherwise, it could result in precipitation if both solutions, i.e. solution A and solution B, are mixed. Solution A and solution B generally come in powder form and are typically poured separately in separate tanks, before being mixed with clean water and stirred.

Once it has become liquid, both A and B will be mixed in another tank which usually holds about 600 litres of water. The AB mixture will be measured using an electrical conductivity (EC) device in order to determine the right amount of fertilizer to be injected into the polybags containing the plants. Aside from ensuring fertilization and irrigation, the farmers would also need to spray insecticide and pesticides as well as other chemical substances to mitigate the threats of insects and fungus which can affect crop growth. Common insecticides would include Abamectin and Imidacloprid that is used to kill thrips and white fly. Similarly, there are specific intervals that need to be observed to successfully suppress the threat of insects and pests. Additionally, the farmers would need to apply foliar to accelerate the growth of the plants.

While fertigation has improved farm management, harvesting activities continue to rely on human labour. Additionally, human labour is needed to conduct regular checks on the water tubes and drippers in case there is a blockage of fertilizers. Therefore, although the fertigation technique offers better plant management, it is still generally laborious. This can be a pressing issue when there is a shortage on the availability of labour. This was especially true during the implementation of the MCO in the early parts of 2020 following the spread of COVID-19.

Automating the Fertigation System

Moving forward, digitalizing the fertigation process could be one of the low-hanging fruits that can be tackled in order to increase crop yield. The present fertigation technique involves laborious work that may result in inaccuracy in many respects.

For instance, the mixture of clean water with AB mixed fertilizers is purely based on manual observation which can be subjected to ambiguity and human error. In these matters, the farmers will generally follow the standard operating procedure (SOP) prescribed by someone like an agricultural consultant. There are no standard competency regulations or licensing for agricultural consultants. Therefore, farmers are left to make critical decisions on their own, especially when faced with weather anomalies that happen during monsoon or drought season. Other emergencies such as electricity outages or a sudden cut in water supplies can also affect the quality of fertilizers received by each plant leading to a lower crop yield.

Such predicaments can be addressed by leveraging the internet of things (IoT). The IoT essentially allows communication between physical devices "the things" with an IT infrastructure or network "the internet". One of the main benefits of IoT is that it will assist the farmers to monitor the operation of their farms which may include monitoring crops and soil conditions to equipment operation and maintenance through dedicated web-based or embedded systems. This system will help farmers to decide on whether to increase or cut the fertigation cycles or watering cycles based on available data. Shah and Noorjannah (2015) have laid out such a framework whereby there are three parts in the proposed system: (1) the web-based system, (2) the automatic fertigation and (3) the communication set-up. Their goal is to create a virtual system that could facilitate farmers in monitoring their farms from their mobile devices as illustrated in Figure 2.5.

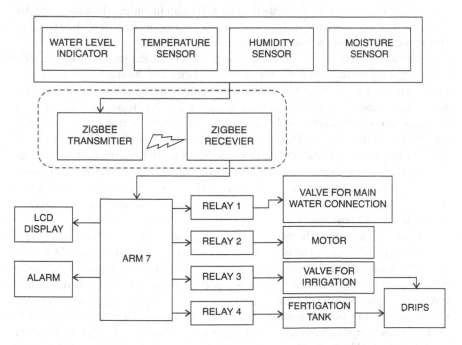

Figure 2.5 Automatic Irrigation and Fertigation using Embedded System.
Source: Shah & Noorjannah (2015).

The web-based systems would allow users to interact with the automatic fertigation system by leveraging the web as the interface or medium. There will be three building blocks for the web which are the database, the web server and the web interface that is usually equipped with a programming language such as SQL, PHP and Javascript. The SQLite database can be used in this project due to the various tools it has for data handling which makes processing easier compared to other process-based (server) relational databases. Such a project can implement a self-contained, serverless, zero-configuration and transactional SQL database engine. The integration between SQLite and application works by using functional and direct calls made to a file holding the data (SQLite database) instead of communicating through interfaces i.e. ports and sockets. The integration happens in the microprocessor where the database resides which can operate faster than other relational databases.

For the webserver, "Lighttpd" (an open-source web-server) is utilized because of its security, speed, compliance and flexibility. Due to the contents of our graphic user interface (GUI), "Lighttpd" is suitable and very efficient because it is designed and optimized for high-performance environments. The web's GUI was designed according to the user's requirement and displays all critical information about the fertigation system. This critical information includes the level of water in tanks and the EC value for the fertilizer solutions. Users are also able to monitor which valve is currently opened in their fertigation system, which could assist farmers virtually in monitoring the flow of the water during irrigation and fertilizers during fertigation. The GUI also includes tabs that show the conditions of the equipment, for example, the status of the valve, the status of the whole system and the flow of pipes.

Users can also remotely operate the automatic fertigation system from the web interface. A queue technique for sorting the priorities of the status is used in this project to avoid any SQL commands accessing the same database simultaneously which can cause database errors. The queue algorithm functions to get data (e.g., the level of fertilizers, the level at mixture tanks, valve status and watchdog for schedule) from the database and used them for web-GUI. The watchdog programme runs in the background for assigned functions. The microcontroller also uses SQL commands to store information from sensors in the database. That is why the queue function is needed for them to take turns in accessing the database. As shown in Figure 2.6, this part consists of a microprocessor, microcontroller and sensors.

The microprocessor is the heart of the system where it is the central processing unit (CPU) between the web interface and the microcontroller. pcDuino, a mini-PC or single-board computer platform that can run operating systems such as Ubuntu and Android ICS, was chosen as the CPU. It has a hardware headers interface compatible with Arduino and can easily be programmed using Python, C and many more [8]. Meanwhile, Mega Arduino is used as the microcontroller to process outputs from sensors. The processed and managed data are then passed to the CPU which is later sent to the database mentioned earlier in part A.

Sensors involved in the automatic fertigation system are ultrasonic sensors, valve sensors and EC sensors that are equipped with a temperature sensor in

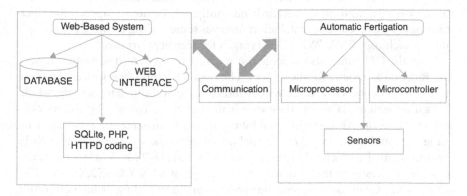

Figure 2.6 Parts of Proposed Automatic Fertigation System.
Source: Shah & Noorjannah (2015).

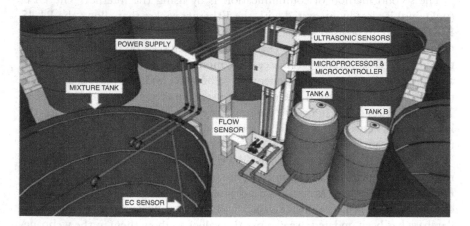

Figure 2.7 Location of the Sensors for Monitoring the Whole System.
Source: Shah & Noorjannah (2015).

order to measure the level of water or solution in a tank, the solution/water from a tank will flow in the pipe connected parallel to it until it reaches the same height as the water in that tank due to the gravitational force exerted in the pipe and tank. The ultrasonic sensors located at the high end of the pipe, as shown in Figure 2.7, examine the height of the pipe from above.

Meanwhile, the valve sensors located in the valve itself will allow the valve to automatically open or close according to the command given. Lastly, the EC sensor is installed in the mixing tank where the EC value of the fertilizer mixture solutions is measured. All the sensors send their feedback in values to the Arduino Mega for data management before they are sent to the database in the CPU. This system can be connected to mobile devices using two methods: either by using a wireless system that is already integrated into the CPU or by using the internet as a medium. The web interface can be accessed from the CPU itself, but it will require having the farmers onsite to perform the action [7]. The wireless network system needs to be installed and configured before the application since it is not pre-installed. The installation involves some coding in the Ubuntu Terminal which uses WPA/WPA2 encryption for security purposes.

Next, the CPU must also become the access point (AP) for the mobile devices and function as a router for the sensors. The reason for this is for dedicated mobile devices to get connected to the system and for users to be able to view the whole fertigation system via a web browser within the range covered by the wireless system. Once the AP has been established, users can manage the system just by using their mobile devices. The channel (range of frequency spread) of the Wi-Fi is set to channel 4 which is in the range of 2414.5 MHz–2439.5 MHz. Next, the number of accesses to the system is set to a range of XXX.XXX.XXX.1 to XXX.XXX.XXX.20. This is to prevent the system from "overloading" which can result in slow access to the wireless system. In this way, users can be around their farm (main office) to monitor it without having to go to their placed system or crops.

The second method of communication is by using the internet. The CPU (pcDuino) is already integrated with the Ethernet socket that allows the RJ-45 Ethernet wire to be used for connection to the internet. However, the IP of the CPU must be set once again to enable the IP to be seen by the main router of the internet. Once set, users can access the monitoring system website anywhere, anytime. Finally, after all components have been set and configured, a verification test must be conducted to ensure that users can access the system anywhere. The verification test is done by "pinging" the IP address of the system for the local wireless network and also keying in the web address of the monitoring system by using any mobile devices for the internet connection network.

The web GUI's of this project are shown in this section. Sensors in the automatic fertigation system as explained in the previous section will send data to the microcontroller (Mega Arduino) for processing and later send to the microprocessor (pcDuino) where updated values will be saved in the database. When the database has been updated, we can use the values to show them in the webpage.

Figure 2.8 shows the overall view of the automatic fertigation system. Tank A and Tank B refer to tank for Fertilizer A and tank for Fertilizer B, respectively. A and B can be any means of fertilizer solution which depends on the type of plants. Tank M refers to the mixture tank where the two fertilizers (A and B) are mixed with water sequentially. The last tank which is the bottom left side is the water tank. The water comes from the water supply body such as SYABAS. The electrical conductivity is monitored in the mixture tank and it is shown under the mixture tank which indicated as "MIXTURE EC VALUE". The EC value is tested on that tank because it is the fertilizer that will flow out to plants.

Figure 2.9 shows the schedule activation page, which allows users to activate any irrigation/fertigation schedule that has already been created in the system. The database for this includes (1) fertilizer's composition (mixture formula) and (2) planting schedule (the day of the plant start to be planted) is set. Users can determine and change parameters of the fertilizers to suit EC of specific crops and create irrigation/fertigation schedule for managing the farm. Once the schedule

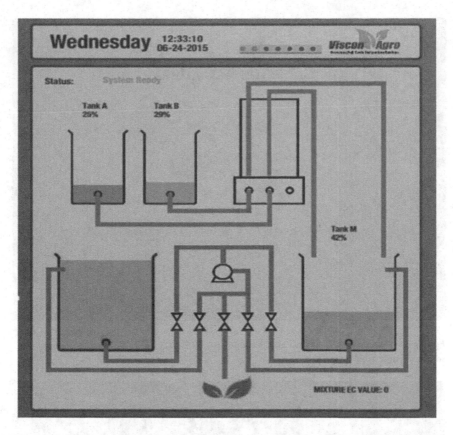

Figure 2.8 Parameters for Monitoring Operations in the Automated Fertigation System.
Source: Shah & Noorjannah (2015)

is activated, this system will run the automatically according to parameters and schedule set in the database. Users also can view the schedule in the Schedule tab.

Figure 2.10 shows the manual mode that allows users to set fertilizer composition manually in for users to choose the ratio of A and B in the case of less or more fertilizer needed for the fertigation process and also the emergency stop button to stop operations. This option may be required when the EC value of the mixture solutions is not achieved within the expected EC value set in the database. The emergency stop button is also in this page where it can be used to halt the system immediately if emergency occurs. Overall functions included in this web-based system are System Overview, List of Schedule, System Log, Settings for Date and time, User Registration, Schedule Activation, Formula Editor, Manual Fertigation and Logout. These features help farmers in managing their farm generally and fertigation system specifically. They can save their time and cost in labour and maintenances.

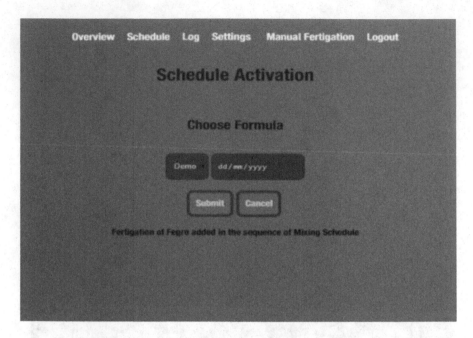

Figure 2.9 Schedule Activation Page, Fertigation System.
Source: Shah & Noorjannah (2015).

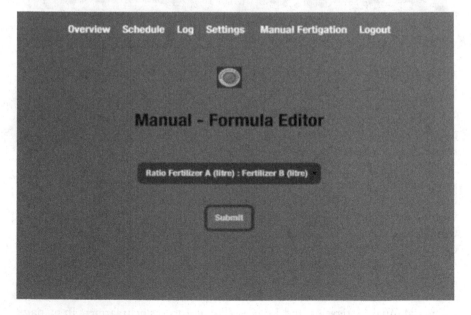

Figure 2.10 Manual Fertigation Page.
Source: Shah & Noorjannah (2015).

Conclusion

The allocation of agricultural lands in Malaysia has increasingly favoured large-scale plantation agriculture. It is very much skewed towards palm oil cultivation, which commands the largest share of arable land acreage in the country. The shift towards large-scale commercial farming has resulted in increased dependence on food imports, which is the major cause of widening of trade deficits in the agro-food sector. The low self-sufficiency rates in many agro-food items suggests that the country is in dire need to revitalize the agricultural sector. Efforts to rejuvenate the food sector inevitably requires the unleashing digitalization and IR4.0 technologies into farming, which is uniquely different in scope-based intensive farming. Consequently, the government's initiatives to quicken the proliferation of digitalization and IR4.0 technologies in food farming should take advantage of the various instruments that have evolved to drive small-scale farming, including the fertigation system discussed in this chapter.

The demand for agro-food products is generally inelastic to prices owing to its peculiar inferior goods' characteristics. As an essential goods industry, there will always be opportunities for business ventures to grow given the large demand-supply gap that exists in the sector. However, the stigma associated with the sector, such as it being a rural industry and the perception that this is not a glamorous industry compared to the oil and gas or the finance sector may discourage local graduates to join the agricultural sector. To this end, digitalizing the sector could offer hopes that the industry would be better accepted among the youths.

Additionally, the fertigation system should involve various other parts, such as water pipes, drippers, valves, water pumps, insecticides, AB mixed fertilizers and foliar. The production of all these inputs can all open opportunities for entrepreneurs to participate in a wide range of clustered businesses. The complimentary rebranding of farming using such IR4.0 technologies through digitalization should also help shift the stigma associated with rural farming.

The fertigation technique has been well-received by farmers based on experts' interactions with government agencies, such as the *Pertubuhan Peladang Kawasan* (PPK; Local Farmers' Organization) in Tanjong Ipoh and Kuala Pilah, which are located in Negeri Sembilan. However, the state of technology in the fertigation system needs a massive leap from largely traditional practices. The high initial cost that would be needed and the uncertainty associated with returns on investment could be the reason for hesitancy among the farmers. Consequently, the initial spur could come from the funds government that have been allocated in the Twelfth Malaysia Plan to modernize small-scale farmers. Based on the channel checks, the cost of setting up a fertigation system could be around RM20,000 for 2,000 poly bags. However, the going rate would differ based on the specification of the design and material used for the system, which is expected to fall as the numbers implementing them reach profitable threshold scale.

Judging from engagement with the critical players in the business, and the relevant government agencies, the traditional fertigation system is still being used as depicted in Figure 2.4. Even awareness of the existing fertigation platform can be deemed as relatively new among the public judging from the social media interaction whereby the traditional fertigation system remains the main topic of discussion. The good thing about social media is that the emergence of fertigation experts has become a major source of information and training for the public. Public training has largely undertaken in a virtual setting, given the restrictions imposed following the outbreak of the COVID-19 pandemic.

To ensure that, the knowledge on the fertigation system, including its setting and operations, should be better regulated in order to ensure its smooth diffusion among the small- and medium-scale farmers. There have been instances whereby the self-proclaimed fertigation experts have taken advantage of participants' ignorance by providing dubious mis-information to the clients. The agro-food sector is also not well-regulated as the players along the value chain, such as the consultants, suppliers and middlemen all operate within their turf. Consequently, farmers are vulnerable to risks, such as cheating, scams and irresponsible conduct, especially among those gullible newcomers to the sector.

The government should also stimulate the quickening of robotization and the introduction of drones in farming, especially in harvesting. It will not only raise precision handling and the elimination of defects and waste, it should also help lessen dependence on foreign labour, and with that lower outflow of remittances, which has had a negative impact on exchange rates. Another option is to increase sourcing labour from the aborigines who are generally more efficient than foreign labour in harvesting chilli. Drastic steps must be taken to quicken the journey towards automating the fertigation system in Malaysia. Apart from that, sourcing local labour is also an arduous task as their commitment to agricultural work is highly questionable. The common solution of procuring foreign labour should be stopped as it not only is expensive but also acts as a deterrent to technological upgrading in the country.

Government support should also increase in handling natural calamities, such as flash floods, protracted droughts and storms, which continues to damage food output in the country. Also, crop devastation from natural disasters has become increasingly unpredictable of late. For example, the flash floods in late 2021 not only accounted for several deaths but also destroyed considerable amount of crops and livestock. The fertigation mechanism will undermined if such calamities are not prevented, which is also critical as some of the agricultural sites are located near rivers. The smooth unaffected operations of fertigation systems require that persistent flash floods do not destroy them. Early digital warning systems, and prompt rescue efforts to insulate the fertigation mechanisms, as well as lives of the people are put in place. A major key area of focus was laid bare by the massive floods that affected Malaysian on 18 December 2021. A key target of the government must be the safety of the 72 dams in Malaysia in 2021. Most of these dams are old.

All signs point to the need for intensifying further the digitalization of the food farming sector. A state-of-the-art fertigation system could translate to higher crop yields as the technology would be able to guarantee the right amount of fertilizer and irrigation goes to the plants. Aside from the immediate impact on the coordinated flow of fertilizers and irrigation, digitalization could also lead to a more detailed orientation towards the collection and use of data (see Rasiah, 2019). The application of big data analytics that help monitor insects, bacteria and viruses could help farmers gauge the developments for farmers to take quick remedial action by applying a more precise amount of insecticide to ward off insects. The current practice is to spray insecticides on a weekly or biweekly arrangement without any coordination with the growth of insects, bacteria, and viruses.

Apart from being meticulous about the types of insects or microorganisms present in the farms, digitalization can also help farmers manage better risk management, which includes anticipating the weather conditions and the probability of occurrence of natural disasters, which will determine the right response to minimize financial losses from such calamities. Overall, digitalization should help alleviate the status of rural farming, offer brighter prospects for income growth and substitute humans with robots and drones in dirty, dangerous and demeaning tasks.

References

CEIC Database (2022). https://info.ceicdata.com/en-products-global-database-ad?utm_campaign=About+Us&utm_source=adwords&utm_term=macroeconomic%20database&utm_medium=ppc&hsa_ad=481036965571&hsa_grp=78296065419&hsa_kw=macro-economic%20database&hsa_ver=3&hsa_net=adwords&hsa_acc=4758588298&hsa_src=g&hsa_cam=6492026826&hsa_mt=b&hsa_tgt=kwd-298263420201&gclid=CjwKCAjwtIaVBhBkEiwAsr7-c3UHOfLpyJyCz8ys8KJ6G4f5PRcOqN95gQ4G3FULvmac0ahfsKJnGBoCFx0QAvD_BwE

DOSM (various issues). *Unpublished Data on Trade Deficits in Agrofood and Self-Sufficiency Ratios*, Putrajaya: Department of Statistics (DOSM), Malaysia.

Malaysia (2011). *The Tenth Malaysia Plan, 2011–2015*, Putrajaya: Government Printers.

Malaysia (2016). *The Eleventh Malaysia Plan, 2016–2020*, Putrajaya: Government Printers.

Rasiah, R. (2018). *Developmental States: Land Schemes, Parastatals and Poverty Alleviation in Malaysia*, Bangi: Universiti Kebangsaan Malaysia Press.

Rasiah, R. (2019). Building Networks to Harness Innovation Synergies: Towards an Open Systems Approach to Sustainable Development, *Journal of Open Innovation: Technology, Markets and Complexity*, 5, 70.

Rasiah, R. & Salih, K. (eds) (2019). *Driving Development: Revisiting Razak's Role in Malaysia's Economic Progress*, Kuala Lumpur: Universiti Malaya Press.

Shah, A.H.Z.A & Noorjannah, S.I. (2015). Web-based Monitoring of an Automated Fertigation System: An IoT Application, Paper presented at the IEEE 12th Malaysia International Conference on Communications (MICC), Kuching, Malaysia, 23 – 25 Nov.

Suresh, P. (2023). Proliferation of IR4.0 Technologies in Large-Scale Agriculture, Rasiah, R., Low, S.W.Y. & Kamaruddin S. (eds) *Digitalizing Development: Ecosystem for Promoting IR4.0 Technologies*, London: Routledge.

3 Proliferation of IR4.0 Technologies in Large-Scale Agriculture

Suresh Palpanaban

Introduction

The agricultural sector has contributed significantly to the growth and development of the Malaysian economy even though it has undergone significant structural changes over the period since 1970. Along with tin mining, agriculture laid the foundation and has been the driving force behind the economic growth of the country. Agricultural proceeds were used to finance the development of the country before export-oriented industrialisation was promoted to spearhead the country's economic growth through export-orientation since the launching of the Second Malaysia Plan in 1971 (Malaysia, 1971).

In 1970, the agricultural sector contributed approximately 31 percent of GDP and employed 50 percent of the labour force. Today, the sector is the third-highest contributor to the national GDP and a key economic sector contributing 8.0 percent in 2020. Palm oil was the major contributor to the value added of agriculture sector at 37.7 percent followed by other agriculture (25.9 percent), livestock (15.3 percent), fishing (12.0 percent), forestry and logging (6.3 percent) and rubber (3.0 percent). While the contribution of the agricultural sector to GDP has declined over the years as the secondary and tertiary sectors have expanded. While such a trend is typical of long-term structural change, which was also followed by the developed economies, the sector has typically recorded positive value-added growth through productivity increments (Rasiah, 2018). This fact was recognised by the government as initiatives were taken through the eleventh Malaysia Plan to drive productivity growth in the sector (Malaysia, 2016). Indeed, in addition to large-scale agriculture, strong emphasis has since been given to small-scale food farming to address the growing dependence on food imports (see Rasiah & Salih, 2019; Afzanizam, 2023; Chamhuri, Rospidah & Mohammed, 2022). The spread of digitalisation and Fourth Industrial Revolution (IR4.0) technologies are among the key strategies advanced by the government to achieve this goal (Malaysia, 2022). As a key economic sector, digitalisation and IR4.0 technologies offer tremendous opportunities to raise the competitiveness and increase productivity in the sector.

Nevertheless, there are significant challenges faced by the sector brought upon by changes in demography and rising population, which according to the UN

DOI: 10.4324/9781003367093-3

Food and Agriculture Organization will add pressure on farmers to generate 70 percent more food output, climate change, crop sustenance and vitality, stress on natural agriculture resources and continued persistence in utilising traditional agriculture practices that impacts food security and sustainability. This inevitably constrains production capacities, disrupts supply chains and requires more food to be produced at greater costs and investments.

Despite the fact that digital and IR4.0 technology adoption provides sufficient demonstrable evidence of increased efficiency and significant transformation in farming practices resulting in higher potential revenue gains and operational effectiveness translating into greater economic advantages, utilisation of IR4.0 technology for greater value add is simply lagging. The purpose of this chapter therefore is to provide pertinent information and reiterate the depth of IR4.0 technology utilisation in agriculture in the context of large-scale farming in Malaysia, the challenges of why this may be underwhelming at this point in time including its propensity and fragmented utilisation uptake, the exacting nature of what is required to elevate its proliferation and measures that are proposed to enable enhancements. In doing so, the chapter focuses on large-scale agriculture in general as there is a separate chapter in the book dealing with small-scale intensive farming (Afzanizam, 2023). The rest of the chapter is organised as follows. The next section presents some theoretical issues to underline the importance of agriculture to most economies followed by the proliferation of IR4.0 technologies in large-scale farming.

Theoretical Considerations

It is a misnomer to think that agriculture shall diminish and disappear as economies undergo a shift towards manufacturing and subsequently services (Rowthorn & Wells, 1987). Given the characteristics of or inferior economic goods, most developed countries have continued to focus on agriculture to avert the occurrence of food import dependency. In the context of this chapter, the focus is not on large-scale food farming. Instead, it is on sustaining agricultural exports with palm oil being arguably one of the rare large-scale sub-sector that carries the characteristics of inferior economic goods as it is an essential edible oil. Among other things, it is because of those reasons, the government introduced export cess to use the revenue collected to both stimulate R&D and stabilise prices (Rasiah, 2022). Following the successful experience of IR4.0 technology application and adoption in agriculture where there has been significant improvements in productivity, reduction in waste and resource utilisation efficiency in several countries, the corollary has been one of several governments promoting its introduction in agriculture (Rasiah, 2018). Whilst acknowledging such nuances, this chapter avoids generalisations by balancing the notion that Malaysia's varied speed of adoption and uptake is largely due to different levels of IR4.0 technology adoption in Malaysia. A one-size-fits-all approach is therefore avoided and in the broader context underscores the point that adoption is but inevitable and urgent.

Along with biotechnology, the focus on sustainable green technologies, IR4.0 technologies have become the cutting-edge drivers of competitiveness since the 1990s, but especially since the turn of the millennium. In Taiwan for example, President Trump's efforts to impose trade restrictions on China drove the relocation of agriculture and manufacturing enterprises back to Taiwan, though the introduction of IR4.0 technologies started earlier. Consequently, robots plough fields and milk cows, while drones spray organic insecticides and fertilisers in Taiwan. In fact, Taiwan has successfully raised rice self-sufficiency rates to exceed 90 percent, and other horticultural products to exceed 80 percent by 2017 (Rasiah, 2018).

Within the IR4.0 technology framework, the most widely used technology pillars include the internet of things (IoT) optimising farm monitoring through smart sensors, unmanned aerial vehicles (UAVs) for soil mapping, imagery and visual aids, agricultural robotics and automation to optimise resource allocation, as well as Big Data for predictive analysis on yields, weather and harvesting cycles (see Rasiah, 2019; Rasiah, Low & Kamaruddin, 2023; Yeap & Rasiah, 2023). Consequently, any attempt to address the policy issues and embedding support environment instruments involving IR4.0 technologies must address the above support instruments along with efforts to create awareness among farmers and the requisite incentive system to quicken the shift to such technologies.

Rasiah (2019) presented an open systems framework that on the one hand discusses the key enablers in ecosystems using the systemic quad model and on the other hand the openness essential to stimulate smartification. Ideally, the linkages between smart farms to intermediary organisations should include the framework presented in Figure 2.1. In addition to the adoption of IR4.0 technologies in large farms, it should include integrating them with the support organisations and mechanisms to evaluate the performance of these linkages in the achievement of desired targets. Such appraisal systems have been effective in the technological transformation of Japan, South Korea, Taiwan and Singapore (Johnson, 1982; Amsden, 1989; Wade, 1990; Rasiah, 2020).

Contribution of Agriculture to Malaysian Economy

The contribution of agriculture to Malaysia's GDP fell in trend terms from 44 percent in 1960 to 7 percent in 2019 before rising slightly to 8 percent in 2020 (Figure 3.1). While such a trend is to be expected owing to the rapid expansion of the secondary and tertiary sectors, the Malaysian experience also shows significant contraction in real value added as in 2000–2001, and 2019. It is plantation farming that has held the sector from contracting sharply. In this regard, much can be done to make large-scale farming (including palm oil), more productive.

The contribution of agriculture's share to Malaysian national employment has also shown a trend fall from 24 percent in 1991 to 13 percent in 2020, albeit the fall has been gentler than in value added share (Figure 3.2). The slower fall in employment contribution is largely a consequence of a slow shift towards automation and IR4.0 technologies. Indeed, the lack of such a transition has also exacerbated Malaysia's dependence on foreign less skilled labour. In fact, restrictions imposed during the COVID-19 pandemic in 2020–2022 has created serious labour shortages in the palm oil sector that it has prevented the farms from taking advantage of rising demand and prices of palm oil since 2021. Consequently, significant amount of palm oil fruits have been rotting on trees (Star, 2022).

In sync with how agriculture has continued to play an important role in medium- and large-developed economies, this chapter emphasises this point to examine why and how digitalisation and IR4.0 technologies should be tapped to stimulate the rejuvenation of the agricultural sector. While Afzanizam (2023) in this book focuses on the small-scale food sector, this chapter addresses the role that institutions and organisations must play in the embedding ecosystem for the expansion of IR4.0 technologies in large-scale farming, including small paddy farms that are operated collectively as large farms.

The relative decline of agriculture, including stagnation and decline in productivity, is a consequence of a number of reasons. For example, the technology used to harvest palm oil fruits have not evolved since the 1970s. While the estate sector is better managed, significant aspects of farming and harvesting have yet to make the transition to cutting-edge technologies. Ploughing fertilising still

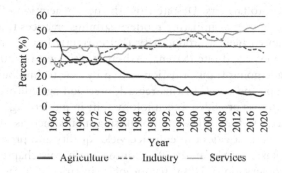

Figure 3.1 Malaysia Gross Domestic Product by Sector, Malaysia, 1960–2020.

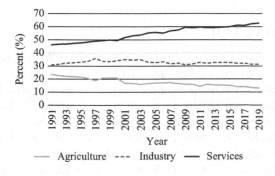

Figure 3.2 Contribution of Agriculture to Employment, Malaysia, 1991–2019.

uses tractors and there is no electronic monitoring of the activities. Watering transitioned from labour-intensive spraying to enabler pipe systems using read only memory (ROM) chips in the 1990s. It has not experienced a shift to drones.

The lack of a comprehensive use of state of the art IR4.0 technologies has produced mixed outcomes in large-scale farming. The prevailing traditional legacy model of farming in Malaysia in general is indicative that the penetration of innovation is low in the sector. One key contributing factor is the ageing famer demography according to statistics from Muda Agricultural Development Authority (MADA) coupled with the lack of willingness to adopt and adapt technology, which is further aggravated by increased urbanisation and over dependence on foreign workers with little or no technology skills raise deliverables' standards (MADA, 2022). The existing use of inefficient technologies has been one major cause of low crop yields, which coupled with wastage due to inadequate farming practices and the lack of IR4.0 technologies has resulted in poor coordination of production and markets. The outbreak of COVID-19 pandemic showed how poorly connected and coordinated are Malaysia's links between farmers and consumers as tonnes of fresh produce by farmers in the Cameron Highlands were dumped.

The situation has been exacerbated by the destruction of irrigation systems, and damming and deforestation that has caused natural disasters with deleterious consequences to farmers. Despite this, the new generation of young farmers are showing signs of assimilating modern farming practices that incorporate technology as an enabler of productivity that currently bridges the gap albeit at a slower than expected rate (Sarena, Ashraf & Siti Aiysyah, 2019). As such, the sector is undergoing transformation driven by new technologies at varying magnitudes, scope and scale. In this regard, precision agriculture (Gebbers & Adamchuk, 2010) has provided a technology shift in modern agriculture. Data is a critical success factor of this and optimising inputs such as nitrates, irrigation, fertilisers and pesticides can enhance yield, quality and productivity (Farhat, Hassan, Aitazaz & Skylar, 2020). Consequently, this chapter attempts to examine the proliferation of IR4.0 technologies in large-scale farming and the steps necessary to strengthen the intermediary organisations in the embedding ecosystem.

Proliferation of IR4.0

In Malaysia, paddy and palm oil have been beneficiaries of precision agriculture under the headline banner of Agriculture 4.0. In order to increase farming efficiency and production systems of these crops, the government has initiated a slew of initiatives to increase the use of emerging technology. These include the use of sensor readings for humidity, temperature, soil moisture, pH, luminance and rainfall. The IoT and video analytics are also widely used applications to measure several variables that impact crop yield, operating cost and inputs such as environmental conditions, growth performance, irrigation, pest and fertilisers, weed management and greenhouse gas emissions. With Agriculture 4.0, technology resources are easily accessible and both the industrial and agro-food commodity industries are adopting IR4.0 technologies through mobile apps, smart sensors, UAVs and cloud computing.

In paddy cultivation, several notable smart farming technologies are evident, such as a geographical information system for monitoring and scheduling daily crop water requirement for paddy (Kamal, & Amin, 2010). Image-sensing technology and UAV application for topological overview of farms is another key utilisation (Tripicchio, Unetti, Giordani, Avizzano, & Satler, 2014). Instead of using humans to investigate, view and determine crop quality and soil integrity, UAVs provide farmers instant real-time images of individual plants for better resource utilisation. Information from those images can be processed to further enhance farmer inputs for optimum yield and crop management.

As with paddy production, the palm oil industry has moved towards adopting a number of elements of smart farming. The smartification processes is already proliferating through automation as a key pillar of the IR4.0 technology adoption with example, such as automated conveyors distributing fruit bunches and weed management. Remote sensors are also employed to support many applications in yield estimation, waste management, automatic tree counting and age estimation (Chong, Kanniah, Pohl & Tan, 2017).

Agriculture 4.0 is also evident in urban farming an example of where vegetables are placed under multi-coloured light-emitting diodes (LEDs) under the Plant Factory with Artificial Lighting scheme that will enable harvesting in a shorter period of time and without the perils of pests and negative impact of pesticides. Another critical area of IR4.0 technology usage comes in the form of livestock farming that consists of automated remote sensors detection and monitoring using individual identification methods that tracks morbidity and mortality indicators using machine learning.

As such, the key element of IR4.0 adoption depends on the level, scope and scale of data analytics, especially for small and medium farmers, which is starting to emerge as a significant enabler to improve the productive capacity, especially in developed counties. The increased access to cloud computing capabilities has enabled the smaller farmer to leverage big data analytics and facilitating a lower barrier to entry for farmers.

Scale and Magnitude of Stakeholder Support and Intervention

Agriculture 4.0 is probably the best platform to transform and increase productivity whilst enabling the shift towards an innovation- and knowledge-based economy. In line with global trends that have witnessed key economic sectors' production capabilities shifting towards IR4.0 technologies, the government's agenda and vision in alleviating poverty, ensuring food security and sustainability through IR4.0 technologies in the sector is gaining traction. The government increased the allocation of the Ministry of Agriculture from RM4.4 billion in 2019 to RM4.9 billion in 2020 as well as provisioning for a special fund for Agriculture 4.0 up to RM43 million. Under the Agriculture Sector Digitalisation and Transformation programme, the government continues to provide support for the industry through various initiatives to transform and empower agro-entrepreneurs, especially in rural areas. One such example would be the shift towards e-commerce platforms within the ambit of digital transformation whereby agro-food entrepreneurs are able to bypass intermediaries to access and penetrate wider markets where online payment mechanisms are used.

The Malaysian Agriculture Research and Development Institute (MARDI) continues to play a pivotal role in providing research in science and technology. MARDI is involved in developing cutting-edge technologies in food processing, post-harvest handling, smart farming technology for paddy and even a public-private strategic collaboration with Maxis to enable precision agriculture for grape crops with narrowband IoT with 5G. This allows farmers to seamlessly scale up the deployment of the solution to wider areas as it has the ability to support millions of connected sensors. Remote monitoring is now possible through the NB-IoT through a web-based app that facilitates alerts to pre-empt negative environmental variables to maximise yield.

Digital AgTech is a pilot initiative driven by the Malaysia Digital Economy Corporation (MDEC) in collaboration with specific partners to empower the agriculture sector by infusing IR4.0 technologies to catalyse digital adoption. Digital AgTech was also recently announced under the 'PAKEJ PERLINDUNGAN RAKYAT DAN PEMULIHAN EKONOMI (PEMULIH)' with KKMM, under the National Recovery Plan. MDEC's efforts to adopt IR4.0 agricultural technology involving various smart farming activities, such as fertigation, fine mist spraying, aquaculture, poultry farming, irrigation and soil monitoring, have also proven successful[9] (https://mdec.my/eladang/).

According to MDEC, innovative fintech and trading platforms such as the blockchain-based peer-to-peer market can be considered to support small farmers by minimising the intermediation process that can provide cost savings to consumers and higher margins for farmers. In palm oil, Felda Global Ventures the commercial arm of Federal Land Development Authority, a Malaysian government agency has invested in automation and mechanisation of daily operations to improve productivity and efficiency deploying advanced technology in geospatial technology applications for oil palm replanting. It produces an accurate replanting blueprint at the oil palm replanting area with automatic palm count information which can later be used as a base map in estimating total fertilisers per hectare, estate management and plant health monitoring. This technology also includes proactive plant health monitoring using precision tools.

The government also made specific provisions for financing the agriculture sector through AgroBank's RM60 million funding for the Agrofood Value Chain Modernisation Programme that offers funding of up to RM1 million at a rate of 3.5 percent for a period of ten year for agricultural entrepreneurs to procure equipment and technology based on IR4.0 technologies. In addition to the above, Table 3.1 depicts a number of initiatives undertaken by multiple stakeholders whether directly or indirectly to enable efficiency improvements in the agriculture sector, including in large-scale farming *(list is not exhaustive)*.

Table 3.1 Initiatives Taken to Raise Efficiency in the Agricultural Sector, Malaysia

Entity	Programs, Schemes
Government agencies and related institutions	
I MARDI	Research and development in agricultural
II Ministry of agriculture and food industries *(RM4.82b allocated for 2022 Budget)*	bio-technology, automation for food processing and engineering technologies as well as providing market access.
III Muda agricultural development authority	Young agropreneur programme through grants which are worth up to RM 20,000,
IV Human resource development corporation	given in the form of machinery, seeds, animal breeds and advisory services.
	Two-day course for young agropreneurs who were taught the intricacies of running a business.
	Industrial Revolution 4.0 scheme under the PENJANA initiative has been providing digital farming training courses.
Financial institutions	
Agrobank's (2022 Allocation of RM1.25 billion financing for Agrobank through among others, Agrofood facility (RM500m) and Dana pembiayaan agromakanan (RM200m)	To provide specific financing facility targeted for young entrepreneur in agriculture sector.
• Program agropreneur muda (PAM);	Financing provision for unemployed graduates who are keen to take up
• Skim Usahawan Tani Komersial Siswazah-i (SUTKS-i)	agriculture.
Commerce international merchant bank's	The fund shall only be used for working capital and/or capital expenditure for
• Agrofood facility scheme	the development of agriculture projects, such as for purchasing machinery and equipment.

Source: MADA, MAFI, HRDF (now rebranded as HRD Corp), Agrobank and CIMB.

The Ministry of Science, Technology and Innovation is in discussions with key stakeholders on policy and regulations related to use of UAVs/systems. This is expected to further enhance current practices that use UAVs for agriculture in paddy, palm oil, banana and pineapple plantations as well as to streamline regulations intra-agency. This also indicates that there is tremendous potential for the use of UAVs as one of the IR4.0 pillars for agriculture in support of full automation (Tsouros, Bibi, & Sarigiannidis, and G., November 2019).

Stakeholder Support Is Not Adequate to Move to the Next Level of Agriculture 4.0

As shown, there are concerted efforts by key stakeholders especially the government in providing a conducive environment for the adoption or assimilation of IR4.0 technologies in large-scale farming in Malaysia, especially in paddy and palm oil which on empirical data shows a moderate uptake scale. Therefore, there needs to be a significant paradigm shift from labour intensive to tech intensive regimes with technology-driven solutions to enhance competitiveness, increase barriers to entry, deliver superior products in terms of price and differentiation by utilising technological know-how in an effective manner to gain competitive advantages. Technology-driven solutions on a broader scale will enable optimisation of resource allocation, minimise operating costs while reducing wastage of raw inputs, efficient farm management, mitigate man-power shortages and deliver greater go to market capacity.

The fact that there is moderate uptake of IR4.0 technologies within the paddy and palm oil sector is largely due to the significant subsidies through government linkages that facilitates investments into related or subsidiary organisations as in the case of palm oil through the Federal Land Development Authority (FELDA) and paddy farmers through the MADA (see also Rasiah & Salih, 2019). While this allows access to catalytic financing assistance, technology acceleration and market penetration, it does not showcase true levels of competitiveness from a technology stand-point.

The competitiveness of the players within the industry is further dampened by a number of other factors, such as:

- *Palm oil* – Ranging from price volatility, anti-palm oil campaigns and protectionist trade regulations in developed countries, particularly in the European Union (EU);
- *Paddy* – Productive capacity and climate change as Malaysia is a net importer of rice. A recent Khazanah Research Institute Report 'The Status of the Paddy and Rice Industry in Malaysia' finds that despite the significant public resources allocated to the industry, paddy farming is still perceived to be uneconomical. This is aggravated further by the fact that the majority of farmers are in the B40 income group and without MADA subsidies cannot remain competitive against key rice growing regions such as Thailand and Vietnam.

The key differences between traditional versus smart farming is, apart from the automation, the manipulation of data in its entirety, including capture, storage, analytics and decision making. While legacy agriculture farming is more visual with windows of subjective decision making, smart farming promotes use of quantitative data for more objective decisions. It is estimated that, with new techniques, the IoT has the potential to increase agricultural productivity by 70 percent by 2050 (Mariani & Kaji, 2016). The IoT, if accurately utilised, will enable the function of big data analytics that will help farmers make informed decisions to generate better revenue opportunities and spill-overs into other tech-related discoveries. It also opens the door to a plethora of massive and valuable information that can be used for the future with predictive analytics tools. However, there still remain challenges for a number of reasons as follows:

Data Interpretation and Utilisation

With massive amounts of data capture for a range of indicators and parameters, that includes amongst others, diagnostics, video imagery of land topology and crops, data from sensors for irrigation and fertilisation, simulation for crop yield prediction, there is a huge trade-off due to the inability for farmers to interpret the type of data collected to run their businesses.

Data Availability, Consistency and Depth

By leveraging IoT, farmers can remotely manage and control their field irrigation systems, monitor soil moisture and crop growth. For example, IR4.0 can offer a weather pattern simulation model to precisely predict oncoming weather so as to enable them to implement mitigation measures. However, not all data can be mined and utilised in the same manner. For example, big data is applicable only in some cases of agriculture and not all farms may have the same intensity of technology adoption. Climate change impact on harvest cycles due to floods and temperature fluctuations create more uncertainty in modelling scenarios for growth targets

Technology Infrastructure and Connectivity

In order to reap the benefits of the IR4.0 technologies, network connectivity is imperative with vital facets such as lower latency, higher bandwidth and increased capacity as key pre-requisites for higher productivity and speed to market with real time information management systems. This, however, is lacking in many areas with prime agriculture development as is the case for paddy farmers in Sabah and Sarawak. While the government's efforts in launching Digital Nasional Berhad to perform the 5G roll-out is commendable, internet access and penetration especially in rural areas where agriculture is a significant economic contributor should be prioritised as an enabler for greater productivity.

COVID-19 Breakout and Impact on Agriculture

During the Movement Control Order as a result of COVID-19, the agriculture industry was one of those that was classified as critical, thus allowing business as usual. But disruptions in the sector was adverse especially due to the labour market crunch, global supply chain and logistical support. There were loss of produce and income and an increase in post-harvest loss and food waste which witnessed farmers dumping fresh produce from Cameron Highlands. One of the single biggest factor of production, labour saw a severe shortage of farm workers specifically in the palm oil industry. The presence of cheap foreign labour is a protagonist for some parts of the agriculture industry in Malaysia but perhaps an antithesis to technology and innovation in Malaysia. This creates an additional stumbling block for the proliferation of a very important IR4.0 technology pillar, automation and robotics which offers promising solutions for Agriculture 4.0 in handling labour shortages and a long-time declining profitability.

Wide Gaps in Technology Innovation

While large-scale farming continues to employ a number of IR4.0 technologies such as smart sensors, IoT and UAVs, existing local capabilities are limited in further tech innovations and discoveries. An example of this would be in drone agriculture. There are a number of drone distributors in Malaysia who supply such platforms for farms but are not themselves a drone integrator. The fact that there is a dearth of available local technopreneurs in drone manufacturing and development amplifies this point. UAVs are basically fixed-wing aircrafts and multirotor aircrafts. In Malaysia, the use of rotary-wing UAVs are more prevalent as they are capable of a vertical take-off and landing but so much slower with a limited battery life vis-à-vis fixed-wing platforms that can cover more area per flight and carry larger payloads. Malaysia is also miles away in utilising autonomous robots in smart farming (Verónica, S.R. and Francisco, 2020) or unmanned guided vehicles as well as in artificial intelligence in the application of machine learning to develop computer programmes that can train actuator/robot to perform a duty.

In this regard, efforts must be taken to expand the deployment of robots and drones in small- and medium-scale agriculture but the whole operations undertaken to appropriate-scale economies through the expansion of collaborative collective farming (see Rasiah, 2018), including palm oil and rubber cultivation. There is a clear need to fabricate robots to harvest fruits, tap trees, plough the fields, water the nurseries and driverless lorries to transport inputs to the farms and output to the refineries. Significant collaboration will be necessary to achieve these developments, including among farmers to appropriate-scale economies through aggregation of acreage, and incubators in science and technology parks to coordinate the development and diffusion of innovation.

Conclusion

While it indeed encouraging to see efforts to introduce IR4.0 technologies in agriculture, the pace of the diffusion needs to gain more speed as the country is still dependent heavily on food imports while much can be done to raise the competitiveness of farms and organisations in Malaysia to enable exports. While digitalisation has been expanding, it is important for the government to offer the awareness, as well as the funds for economic agents engaged in large-scale farming to acquire and absorb IR4.0 technologies to raise productivity while reducing wastage. The transformation to IR4.0 technologies should also help the government to reduce dependence on foreign workers in the country.

In view of current challenges related to food security, climate change and global warming, modernisation and depletion of agriculture land, innovation and technology adoption have emerged as key tools to enable the evolution of sustainable agricultural productive capacity. Failure to embrace the changes rapidly brought upon by technological advancements can severely impact economic growth and social well-being of the population, which can seriously aggravate further Malaysia's heavy reliance on imports of key food inputs. The further the disconnect between access to and adoption of technological innovations in large-scale farming, the greater the gap between production and mid-stream players that could result in severe disruptions to the supply chain. Therefore, there is an urgent need to adopt rapidly and integrate key IR4.0 technologies to enable significant cost and yield advantages. While the government continues to provide increased support for the agriculture sector through efficient infrastructure, greater incentives including grants, financial inclusion and access to financing, the production itself must continue to aggressively innovate and embrace deep tech from current low adoption but promising streams to highly competitive large-scale farming centres.

References

Afzanizam, M.A.R. (2023) Improving Food Production through Digitalizing the Fertigation System, Rasiah, R., Low, S.W.Y. & Kamaruddin S. (eds), *Digitalization and Development: The Ecosystem for Promoting Industrial Revolution 4.0 Technologies in Malaysia* , London: Routledge.

Amsden, A.H. (1989) *Asia's Next Giant: South Korea and Late Industrialization*, New York: Oxford University Press.

Chamhuri, S., Rospidah, G. & Mohammed, I. (2022) Agriculture and Food Security, Rasiah, R., Salih, K. & Cheong, K.C. (eds), *Malaysia's Leap Into the Future: The Building Blocks Towards Balanced Development*, Singapore: Springer (pp 117–139).

Chong, K.L., Kanniah, K.D., Pohl, C. & Tan, K.P. (2017a) A Review of Remote Sensing Applications for Oil Palm Studies, *Geo-Spatial Information Science, 20*(2), 184–200.

Chong, K.L., Kanniah, K.D., Pohl, C & Tan, K.P. (2017b). A Review of Remote Sensing Applications for Oil Palm Studies. *Geo-Spatial Information Science, 20*(2), 184–200. https://mdec.my/eladang/

Farhat, A., Hassan, A., Aitazaz, A.F. & Skylar, T. (2020) Crop Yield Prediction Through Proximal Sensing and Machine Learning Algorithms, *Agronomy, 10*(7), 1046.

Gebbers, R. & Adamchuk, V. (2010) Precision Agriculture and Food Security, *Science, 327*(5967), 828–831.

Johnson, C. (1982) *MITI and the Japanese Miracle: The Growth of Industrial Policy, 1925–1975*, Stanford: Stanford University Press.

Kamal, R.M. & Amin, M.S.M. (2010) GIS-based Irrigation Water Management for Precision Farming of Rice, *International Journal of Agricultural and Biological Engineering, 3*(3), 27–35.

MADA (2022) Make Agriculture Attractive Among Youths, *Malaysian Reserve*, June 13, accessed on June 13, 2022 from https://themalaysianreserve.com/2021/06/23/make-agriculture-attractive-among-youths/

Malaysia (1971) *Second Malaysia Plan 1971–1975*, Kuala Lumpur: Government Printers.

Malaysia (2016) *Eleventh Malaysia Plan 2016–2020*, Putrajaya: Government Printers.

Mariani, J. & Kaji, J. (2016). From Dirt to Data: The Second Green Revolution and IoT, *Deloitte Review Issue* 18, accessed on June 13, 2022 from https://www2.deloitte.com/us/en/insights/deloitte-review/issue-18/second-green-revolution-and-internet-of-things.html

Rasiah, R. (2018) *Developmental States: Land Schemes, Parastatals and Poverty Alleviation in Malaysia*, Bangi: Universiti Kebangsaan Malaysia Press.

Rasiah, R. (2019) Building Networks to Harness Innovation Synergies: Towards an Open Systems Approach to Sustainable Development, *Journal of Open Innovation: Technology, Markets and Complexity*, 5, 70.

Rasiah, R. (2020) Industrial Policy and Industrialization, Arkebe, O., Cramer, C., Chang, H.J., & Khozul-Wright, R. (eds), *Oxford Handbook of Industrial Policy*, Oxford: Oxford University Press.

Rasiah, R. (2022) Manufacturing Transformation, Rasiah, R., Salih, K. & Cheong, K.C. (eds), *Malaysia's Leap Into the Future: The Building Blocks Towards Balanced Development*, Singapore: Springer, pp 141–164.

Rasiah, R., Low, S.W.Y. & Kamaruddin, S. (eds) (2023) *Digitalization and Development: The Ecosystem for Promoting Industrial Revolution 4.0 Technologies in Malaysia*, (cross referenced from same book).

Rasiah, R. & Salih, K. (2019) *Driving Development: Revisiting Razak's Role in Malaysia's Economic Progress*, Kuala Lumpur: Universiti Malaya Press.

Rowthorn, R. & Wells, J. (1987) *De-industrialization and Foreign Trade*, Cambridge: Cambridge University Press.

Sarena, C.O., Ashraf, S. & Siti Aysyaf, T. (2019) *The Status of the Paddy and Rice Industry in Malaysia*, Kuala Lumpur: Khazanah Research Institute.

Star (2022) Malaysia Unable to Cash in on Rising Oil Palm Prices, *Star*, May 11, accessed on May 23, 2022 from https://www.thestar.com.my/news/nation/2022/05/11/msia-unable-to-cash-in-on-rising-palm-oil-prices?fbclid=IwAR3YX7AdM-7SUP6xU-pXhjQCR0YfhjdmJ3UpODq4i9SDNBFlMl25IGjGm8KE

Tripicchio, P., Unetti, M., Giordani, N., Avizzano, C.A. & Satler, M. (2014) A Lightweight Simultaneous Localization and Mapping Algorithm for Indoor Autonomous Navigation. In *Proceedings of the 2014 Australasian Conference on Robotics and Automation (ACRA 2014)*, 2–4.

Tsouros, D.C., Bibi, S. & Sarigiannidis, P.G. (2019) A Review on UAV-Based Applications for Precision Agriculture, *Information*, 10(11), 349.

Verónica, S.R. & Francisco, R.M. (2020) From Smart Farming towards Agriculture 5.0: A Review on Crop Data Management, *Agronomy*, 10, 207.

Wade, R. (1990) *Governing the Market: Economic Theory and the Role of Government in East Asian Industrialization*, Princeton: Princeton University Press.

Yeap, K.L. & Rasiah, R. (2023) Diffusion of IR 4.0 Technologies in Electronics Manufacturing: The Role of the Embedding Ecosystem, Rasiah, R., Low, S.W.Y. & Kamaruddin, S. (eds), *Digitalization and Development: The Ecosystem for Promoting Industrial Revolution 4.0 Technologies in Malaysia*, London: Routledge.

4 Diffusion of IR4.0 Technologies in Electronics Manufacturing

The Role of the Embedding Ecosystem

Yeap Khai Leang and Rajah Rasiah

Introduction

The concept of Industry 4.0 was conceived in 2011 by the German government with the intention of introducing a new concept for German economic policies that utilise high-tech strategies (Mosconi, 2015). It is the driving component of the Fourth Technological Revolution combining the usage of cyber-physical systems (CPS), the internet of things (IoT), the internet of services (IoS), and smart factories (Lasi et al., 2014). The phenomenon relies on communication through the internet for constant interaction and information exchange between humans and machines (Cooper and James, 2009). It is considered a vision that is "based on a network of autonomous, self-controlling, self-configuring, knowledge-based, sensor based, and spatially distributed production resources" (Wilkesmann and Wilkesmann, 2018). Industry 4.0 will have a tremendous impact on the digital industry in three ways: it will encourage the digitisation of production, complete automation, and linkages throughout a supply chain. The nine pillars that define the process are big data and analytics, autonomous robots, simulation, horizontal and vertical system integration, industrial IoT, cyber security and cyber-physical systems, the cloud, augmented reality, and additive manufacturing (Vaidya, 2018).

Digital infrastructure and analytical capabilities from the Third Industrial Revolution (IR3) provided the foundation for IR4.0 technologies (Castelo-Branco et al., 2019). Industry 4.0 added its innovative capabilities to operational processes across all types of manufacturing supply chains, a variety of industries, and firms of any size (Wan et al., 2017). The concept's ultimate purpose is to create agile companies that have the capability to continuously learn and adapt, in accordance with a dynamic changing market environment (Schuh et al., 2017). Its implementation in manufacturing systems anticipates a more efficient, flexible, sustainable, higher-quality and lower-cost form of production (Wang et al., 2015). Scholars have attempted to research the application and effects of Industry 4.0 in specific countries. Wilkesmann and Wilkesmann (2018) concluded that firms in Germany have experienced a variety of outcomes differing in the degrees of formalisation, location of control authority, location of knowledge, and degree of professionalism. Beyond its application in Germany, several other

DOI: 10.4324/9781003367093-4

nations have attempted to embrace the Industry 4.0 strategy, including the United States, China and Malaysia, amongst others (Zhong, 2017).

The implications of Industry 4.0 on the social and economic landscape are both unprecedented and uncertain. The implementation has the potential to alter the everyday lives of the general population through effects on the labour market, educational preparation, consumer experience, and more. The objective of this research is to understand the industry-specific ecosystem variables that influence the diffusion of Industry 4.0-related technologies and how they can be enhanced.

The E&E sector was chosen because it is a critical contributor to the Malaysian economy, which along with aerospace is often considered technologically the most advanced industry in Malaysia. Consequently, it serves as an ideal testbed for further development through targeted policymaking and better institutional synergy for a country seeking to deepen its engagement in high technology industrialisation (Rasiah, 2020). A firm-level survey of 50 E&E manufacturers in Malaysia was carried out in 2019–2020 for the purpose of this research with a focus on the role of embedding institutions and organisations in the introduction of IR4.0 technologies.

This chapter is structured as follows. Following the introduction, Section "Theoretical Considerations" consists of a literature review that addresses the critical theoretical arguments relevant to the study, with a focus on existing studies that have broached the impact of the ecosystem impact on the proliferation of Industry 4.0 technologies in the E&E sector. Section "Methodology and Data" introduces the research methodology undertaken, including the empirical data collection method and sample of interviews conducted. Section "Findings and Analysis" discusses the findings. Finally, Section "Conclusions" presents the conclusion.

Theoretical Considerations

This section examines the extant literature on the origin and proliferation of IR4.0 technologies in the electric and electronics industry to locate the significance of the ecosystem supporting the evolution and functioning of such technologies.

Origin of Industry 4.0

The First Industrial Revolution is said to have taken place over the period of 1712 to 1912 when steam engines were introduced in Western Europe, allowing machine-based production. The Second Industrial Revolution was embraced through the usage of electrical energy for mass production purposes up until 1968, which emerged with the new assembly line systems, including the Taylorist and Fordist versions. Between 1969 and 2012, electronic devices and information technologies were integrated, which drove increased automation in industrial manufacturing activities (Bauer et al., 2018; Hänninen et al., 2018).

Through computerisation, the IR3 provided the foundation for the emergence of Industry 4.0 in manufacturing (Schuh et al., 2017). These prerequisites include digital infrastructure and analytical capabilities (Castelo-Branco et al., 2019) Further preconditions mentioned by Rojko (2017) are production stability during systems transition, availability of continuous planned financial investments, and knowledge of cyber protection.

Industry 4.0 took shape at the turn of the millennium as artificial intelligence drove the introduction of robots and drones as German researchers, Henning Kagermann, Wolf Dieter Lukas, and Wolfgang Wahlster conceptualised it in 2011 at the Hannover Fair, which focused on a new economic policy that integrated hi-tech strategy to strengthen the competitiveness of the German manufacturing firms (Kagermann et al., 2013; Mosconi, 2015). Industry 4.0 gradually then became the foundation of a national initiative called "High-Tech Strategy 2020 for Germany," followed by the assembly of a designated working group that took on suggestions for implementation (Kagermann et al., 2013: 5).

Critical Attributes

Specifically, IR4.0 can be differentiated from earlier industrial revolutions by three features, *viz.*, increasing automation and digitisation (Brettel et al., 2014; Oesterreich and Teuteberg, 2016), the agglomeration of a collective group of technologies (Hermann et al, 2016; Pfohl et al, 2017; Szalavetz, 2019), and outcomes produced by the transformation expected in production (Kagermann et al, 2013; Pereira & Romero, 2017). Consequently, Industry 4.0 activities can be considered as the accumulation of the processes, technologies, and outcomes that derive from the intended digital transformation of the industry.

In manufacturing, Industry 4.0, the introduction of artificial intelligence, provides innovative capabilities to operational processes across all types of supply chains, industries, and firms of any size (Wan et al., 2017). The concept has the ultimate purpose of creating agile companies that have the capability to continuously learn and adapt, in accordance with a dynamic changing market environment (Schuh et al., 2017). Its implementation in manufacturing systems anticipates a more efficient, flexible, sustainable, higher-quality, and lower-cost form of production (Wang et al., 2015). Beyond the technical elements, Industry 4.0 is expected to create changes in the relationships between organisations among employees, value chains, nature, and local communities (Santos et al., 2017).

Industry 4.0 drives changes in horizontal integration, vertical integration and end-to-end engineering integration (Kagermann et al., 2013; Wang et al., 2015). Horizontal integration encourages cooperation between corporations within the same value creation networks to form efficient ecosystems through resource and information exchange. By doing this, these collective enterprises can combine resources, share risks, and adapt to changes in markets to access new opportunities (Brettel et al., 2014). This is the overall connection of value creation modules within their individual systems (Stock et al., 2018).

Vertical integration is the integration of information communication technology (ICT) systems at separate hierarchical levels within an organisation that allows the integration of production and management systems. The sub-systems utilised across all processes within an organisation must be vertically integrated so that smart machines can continuously self-adapt to serve different production purposes and continuously relay relevant data for self-improvement and enhanced transparency. This mainly contributes to the dynamic and reconfigurable abilities of the manufacturing systems allowing self-controlling systems (Stock et al., 2018; Wang et al., 2015). End-to-end engineering integration is the value creation process that focuses on the entire value chain of a product. It is the connection created between all phases of a product's life cycle, from development up until post-sales (Gilchrist, 2016). By doing this, the product model is consistent throughout each stage and customisation is made feasible (Wang et al, 2015). The meta-integration of all three processes allows full integration at every stage of the product life cycle through a flexible manufacturing system (Foidl and Felderer, 2016).

Components

The interrelated nature of manufacturing, in general, and an innovation-driven industry (such as electronics) have resulted in difficulties in clearly identifying and categorising the technologies involved. Chiarello (2018) mentioned the four main challenges: a large number of constituent technologies, understanding and usage of these technologies are dependent on the circumstantial application, the existence of a variety of stakeholders, and difficulty of constantly adapting in accordance with rapidly evolving technical progress. Considering this, Industry 4.0 technologies have been grouped in several ways in the literature. The most generic grouping differentiates the technologies into a physical process-driven technologies and digital technologies. Physical technologies are mostly hardware components utilised for the improvement of production processes in the physical sense, including additive manufacturing, collaborative robots, and transport systems (Gibson et al., 2014).

Digital technologies can be further broken down into three sub-groups: digital interface technologies that connect physical systems with cyber ones, network technologies that allow online functional capabilities, and data processing technologies that conduct analysis with collected information for decision-making processes (Culot et al., 2020). Digital interface technologies are mostly hardware components with the element of network connectivity like cyber-physical systems (CPS), IoT, and visualisation technologies (Chryssolouris et al., 2009; Lee et al., 2015; Lee and Lee, 2015). Network technologies are mostly software components like cloud computing and cybersecurity solutions (Armbrust et al., 2010; Anderln, 2015). Finally, data processing technologies are software components that are information-driven like artificial intelligence, simulation, and big data analytics (Chen et al., 2014; Yao et al., 2017; Brödner, 2018).

Acknowledging the above, there exists a collection of relevant technologies that have been commonly referred to as the nine pillars of Industry 4.0 technologies, *viz.*, big data analytics, autonomous robots, simulation, horizontal and vertical systems integration, IoT, cybersecurity, cloud computing, additive manufacturing, and augmented reality (Rüßmann et al., 2015). The variations in the interpretation of the enabling technologies usually derive from a mix of the nine pillars. Frank et al. (2019) prioritised connectivity and intelligence capabilities of just the IoT, cloud computing, and big data analytics while Rojko (2017) puts a strong emphasis on the application of CPS and IoT. Moeuf et al. (2019) neglect additive manufacturing, and replace it with CPS, while replacing systems integration with machine-to-machine (M2M) communication, and altering autonomous robots to collaborative robots. Some authors narrow down the scope to just four major technological components, *viz.*, CPS, IoT, and IOS, and smart factories (Hermann et al, 2016; Roblek et al., 2016).

For the purpose of this research, 17 technologies are observed. These are divided into five technology groups: infrastructure, tracking and monitoring, production, transport, and digital. Infrastructure technologies refer to cloud computing and big data collection. Tracking and monitoring technologies include digitalised control systems, real-time tracking technologies, sensors, radio frequency identification, and M2M communication. Production technologies are 3D printing and additive manufacturing. Transport technologies comprise autonomous robots and drones. Finally, cyber or digital technologies focus on the IoT, big data analytics, cybersecurity, simulation, augmented reality, and artificial intelligence.

Developments in Manufacturing

Due to the complicated architecture of Industry 4.0 systems and its relatively recent emergence, the implementation of its associated technologies is still an active subject of research (Lee et al., 2015; Babiceanu and Seker, 2016; Dalenogare et al., 2018). There are limited numbers of practical studies utilising either quantitative or qualitative methods to measure the phenomenon (Piccarozzi et al., 2018). The types of Industry 4.0 analyses methods that have been undertaken in previous literature include implementation frameworks, technological roadmaps, maturity models, technological cluster models, readiness assessments, policy reviews, and empirical surveys.

Several authors have carried out empirical studies that utilised similar methods to collect responses from firms regarding their levels of Industry 4.0 technology usage to determine adoption patterns. Chiarini et al. (2020) carried out a survey among 200 Italian manufacturers to measure their degree of implementation and consideration for the adoption of Industry 4.0 technologies. Rauch et al. (2020) conducted a study with 17 companies across four countries to validate their SME maturity level-based assessment tool on the implementation of Industry 4.0. Benitez et al. (2020) attempted to explore the evolution of the Industry 4.0 ecosystem in Brazilian electrical and electronics companies through 37 interviews with stakeholders and firms across several years. Pessot et al. (2020) investigated the implementation of Factory of the Future concepts and digitalisation in 92 manufacturing firms in the Alpine regions through a qualitative survey and in-depth interviews. Frank et al. (2019) looked into organising firms into Industry 4.0 technological adoption patterns. With 92 respondents, they designed an adoption framework for the specific technologies decided upon by a group of supposed experts. Sundberg et al. (2019) assessed the digital maturity of manufacturing companies in Sweden through the distribution of digital surveys to 123 firms. Vrchota et al. (2019) utilised a mixed method research with a survey study of 1018 SMEs and case studies of 72 SMEs in the Czech Republic. Calabrese et al. (2020) interviewed 39 managers in multiple sectors to understand the adoption rates of enabling technologies in relation to firm size using an E-SWOT analysis. None of these previous studies considered the embedded ecosystem and its role in encouraging the diffusion of Industry 4.0.

Ecosystems to Promote Industry 4.0

Considering the limited studies that evaluate the relationship between firms' ecosystem and Industry 4.0 adoption, the literature has been explored through the theoretical lenses of technological capabilities and digitalisation. A firm's technological innovation environment can be separated into internal and external parts that influence its capabilities (Berry et al., 2010). A significant external environmental factor for technological capabilities is the different ways governments provide support for firms (Damanpour, 1991). Industrial policy exponents, including Hamilton (1791), List (1885), and Smith (1776) highlighted the importance of institutional support for innovation. Coase (1937), Williamson (1985), and North (1990) regarded institutions as the structures and rule-makers that govern economic activity and firm performance. Veblen (1915) and Nelson and Winter (1982) argued over the social institutions that govern physical technological development and diffusion. Rasiah, Rafat, and Yap (2016) confirmed that host-site institutions and other meso-organisations that support the appropriation of knowledge play a big role in how firms acquire technological capabilities. The existing literature that studies the advancement in digitalisation indicates that the embedding ecosystem of a firm plays a vital role in its technological development. Industry 4.0 is similar in nature and introduces new technological capabilities to the firm, thus the embedded ecosystem of the firm is presumed to have an influence.

States are the most important stakeholders when it comes to strengthening the embedded ecosystem. In support of firms' technological development, the government's role begins with providing suitable basic and digital infrastructure, as well as providing incentives and opportunities for increasing firm competitiveness. Depending on each country, states can choose to play an even more active role in the stimulation of advanced technology usage through building synergies and developing government-linked support organisations. Governments can also prioritise strong policies regarding intellectual property protection and cybersecurity systems. Investing in human capital development is another way that states can ensure a strong ecosystem for firms, this is conducted through revising syllabus and setting standards within educational and training institutes that are aligned with the changing developments of technologies. A highly skilled labour force is ideal for increasing productivity and potential commercial gains from firms.

Rasiah (2019) had examined the role of embedding ecosystem using the systemic quad, which among others looked at the role of broadband infrastructure within basic infrastructure, and the science, technology, and innovation infrastructure that is critical to establishing systemic knowledge networks, and to support connectivity and coordination between individuals, firms and supporting organisations. This paper discussed the role of knowledge networks to raise smartification synergies among a cluster of firms in particular locations. In addition, it discussed the need for transforming knowledge to open systems sourcing owing to the critical importance of such channels for appropriating societal synergies where it involves public utilities, such as power, health, and water, and public goods, such as knowledge and the environment.

Wider national policies through frameworks and roadmaps targeting digital infrastructure building are considered critical to promoting the proliferation of IR4.0 technologies. Malaysia launched its own IR4.0 Master Plan in 2018 led by the Ministry of International Trade and Industry and later, the Malaysian Digital Economy Blueprint in 2020. These are positive initiatives that highlight the importance of this newer wave of production technologies for firms and the increasingly active role that the government would like to play in encouraging its adoption. There continue to be opportunities to further develop the ecosystem for enhancing the diffusion of Industry 4.0 amongst manufacturing firms in Malaysia.

Having examined the extant literature on the empirical evidence and ecosystems stimulating IR4.0 in electric and electronics manufacturing, the next section will introduce the methodology to examine how the related ecosystem has evolved in Malaysia. This research will investigate the embedding ecosystem through firm perceptions of the quality of existing institutions, policies, organisations, and stakeholder relationships. Industry 4.0 research that studies the potential improvements for ecosystems is still lacking and there is a need to further understand how this links with firm technological adoption.

Methodology and Data

The research methodology used in this chapter can be broken down into two components: primary research using a research design to collect information on firms' assessment of the eco-system facing them, and a quantitative analytic framework. A review was conducted on methodologies used in past works focused on supporting organisations and instruments for Industry 4.0 adoption and associated technologies.

Questionnaire and Sampling

A questionnaire was developed by adapting the one used by Rasiah and Yap (2015a, 2015b) by using Likert scale scores to capture electric and electronics firms' (E&E) rating of critical supporting organisations that are important for the promotion of Industry 4.0 technologies.[1] Unlike the questionnaires used by Rasiah and Yap (2015a, 2015b), the questionnaire used for this study was specifically targeted at Industry 4.0 technologies. The population used for this study was defined by the largest database available from the Malaysian Industrial and Development Authority (MIDA). This list consisted of 485 companies across Malaysia that have been engaged in all the sub-sectors of the E&E industry. A stratified random sampling procedure based on size and ownership was originally adopted to gather data from the E&E sector in Malaysia. However, owing to problems of access, the stratified sample was eventually transformed following a pilot study on ten firms that was recommended by the industry association. Consequently, a non-discriminative snowball sampling method was eventually deployed from which 55 companies were selected for the purpose of this research. These firms were directly approached and eventually, 50 firms joined the survey, delegating one top manager who was directly involved with the production process for a face-to-face interview that lasted between 45 and 90 minutes each. The response rate was 90.9 percent of the sample of firms identified. All personal and company details were kept anonymous to ensure non-bias responses from the employees, as well as to meet the confidentiality guarantee assured to the respondents, and the ethics committee agreement.[2] Tables 4.1–4.4 are summaries of some basic firm characteristics of the sample, including each firm's respective E&E subsector, ownership, size, and location by state.

Table 4.1 Respondent Firms by Malaysian E&E Sub-sectors, 2020–2021

Firm Sub-sectors	Number of Respondents
Electronic components	26
Consumer electronics	2
Industrial electronics	6
Electrical appliances	7
Electronic machinery and equipment	9
Total	50

Source: Authors survey.

Table 4.2 Respondent Firms by Ownership, 2020–2021

Firm Ownership	Number of Respondents
Foreign owned	33
Nationally owned	17
Total	50

Source: Authors' survey.

Table 4.3 Respondent Firms by Size, 2020–2021

Firm Size (Number of Employees)	Number of Respondents
Less than 100	9
100 to 499	6
500 to 999	9
1000 to 2999	15
More than 3000	11
Total	50

Source: Authors' survey.

Table 4.4 Respondent Firms by Location, 2020–2021

Firm Location (by State)	Number of Respondents
Johor	3
Kedah	3
Melaka	1
Negeri Sembilan	1
Penang	30
Perak	1
Sarawak	1
Selangor	10
Total	50

Source: Authors' survey.

A cross-sectional survey was administered through a face-to -face semi-structured interview guided by a specific questionnaire instrument designed by the authors. For firms approached during the restricted movement control period, the survey was administered through an online platform and the interview had been conducted virtually. Content validity of the survey instrument was assured through a pilot test where the authors interviewed ten companies from the E&E sector and five experts. Out of the ten companies, seven were eventually included in the actual study as technology users, while three were considered technology providers. Consultations with five experts from academia, government, business associations, and industry representatives also helped to shape the survey instrument.

Questions from the survey instrument were used to measure the state of process technology adoption, with an emphasis on the usage intensity of the Industry 4.0 technologies identified. Additionally, the survey assesses the respondent's perceptions of the ecosystem that supports Industry 4.0 adoption. All factors were measured using Likert scale scores (1–5).

Analytical Framework

The analytical framework constructed for this chapter is shown in Figure 4.1. The different instruments and organisations associated with the adoption of IR4.0

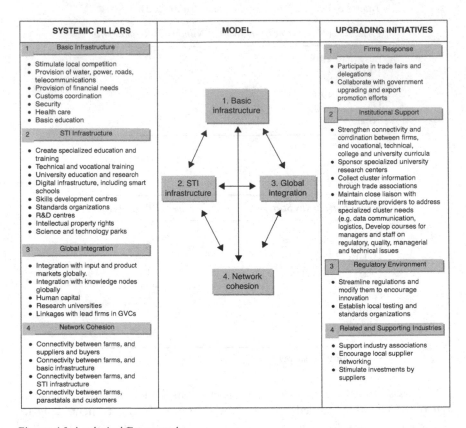

SYSTEMIC PILLARS	MODEL	UPGRADING INITIATIVES

1 Basic Infrastructure

- Stimulate local competition
- Provision of water, power, roads, telecommunications
- Provision of financial needs
- Customs coordination
- Security
- Health care
- Basic education

2 STI Infrastructure

- Create specialized education and training
- Technical and vocational training
- University education and research
- Digital infrastructure, including smart schools
- Skills development centres
- Standards organizations
- R&D centres
- Intellectual property rights
- Science and technology parks

3 Global Integration

- Integration with input and product markets globally.
- Integration with knowledge nodes globally
- Human capital
- Research universities
- Linkages with lead firms in GVCs

4 Network Cohesion

- Connectivity between farms, and suppliers and buyers
- Connectivity between farms, and basic infrastructure
- Connectivity between farms, and STI infrastructure
- Connectivity between farms, parastatals and customers

MODEL:
1. Basic infrastructure
2. STI infrastructure
3. Global integration
4. Network cohesion

1 Firms Response

- Participate in trade fairs and delegations
- Collaborate with government upgrading and export promotion efforts

2 Institutional Support

- Strengthen connectivity and cordination between firms, and vocational, technical, college and university curricula
- Sponsor specialized university research centers
- Collect cluster information through trade associations
- Maintain close liaison with infrastructure providers to address specialized cluster needs (e.g. data communication, logistics, Develop courses for managers and staff on regulatory, quality, managerial and technical issues

3 Regulatory Environment

- Streamline regulations and modify them to encourage innovation
- Establish local testing and standards organizations

4 Related and Supporting Industries

- Support industry associations
- Encourage local supplier networking
- Stimulate investments by suppliers

Figure 4.1 Analytical Framework.
Source: Rasiah (2019).

technologies in the embedding ecosystem are studied to assess their current state of development and their effects on firm-level technological development and Industry 4.0 adoption in Malaysia's E&E manufacturing sector. In doing so, the institutions and intermediary organisations evaluated by the firms include the critical ones identified in the systemic quad developed by Rasiah (2019), i.e., basic infrastructure, science, technology, innovation infrastructure, global integration, and network cohesion (Figure 4.1). The distinctive elements that differentiate basic and science, technology, and innovation infrastructure are that the former is largely constituted of public utilities that are excludable and rivalrous while the latter is constituted by public goods that are non-excludable and non-rivalrous.

Consequently, this chapter focuses on the E&E firms' rating of the organisations and institutional instruments of the four critical pillars of the ecosystem embedding them in Malaysia.

Findings and Analysis

The findings from this study are interpreted in the four sub-sections below. First, we seek to understand the overall embedded ecosystem of the firm by assessing the variables that affect their operational functions and technological adoption. Second, we further investigate the quality domestic environment influences. Third, the focus is placed on evaluating the quality of government-linked institutions and other forms of support. Finally, the external stakeholders and the value of each relationship with the firm are considered.

Quality of Infrastructure and Organisations

For infrastructure and organisations, respondents were asked to rate the quality of basic utilities including the systems for water, power, transport, and internet broadband; public institutions like health facilities, primary education, and universities; financial access to capital or credit; basic government functions; and the scientific and technological ecosystem through training institutions, R&D scientists and engineers, incentives for R&D, and R&D specific institutions.

Table 4.5 ranks the average response for each variable assessed using Likert scale scores. The scores range between twos and threes indicating that the overall sentiment about the quality of infrastructure and organisations is neutral or below average levels of satisfaction. Basic utilities and digital infrastructure are the highest rated, whereas the science and technology-specific ecosystem factors have been rated considerably poorly. The quality of educational, research, and training institutions is perceived to be very low amongst the responding firms – especially the R&D scientists and engineers, incentives provided for R&D, and R&D specific institutions. This suggests that there is a generally poor opinion of the availability and reliability of R&D specific support and its adjacent labour force.

Table 4.5 Quality of Infrastructure and Organisations in Malaysia, 2020–2021

Infrastructure and Organisations	*Mean Response*
Water supply	3.86
Power supply	3.7
Public health facilities	3.46
Access to capital/credit	3.34
Internet Broadband	3.16
Transport systems	3.02
Coordination from basic government agencies	2.98
University education	2.86
Primary schools	2.84
Technical training institutions	2.84
R&D scientists and engineers	2.48
Incentives for R&D	2.48
R&D institutions	2.24

Source: Authors' survey.

Domestic Environment for Technological Development

To assess the quality of the Malaysian domestic environment for technological development, respondents evaluated the government incentives provided for innovation, supply of skilled and scientific manpower, availability of compatible local universities and quality R&D institutions for technical collaboration, intellectual property protection, quality of ICT services in the domestic market, and the availability of alternative funding including venture capital institutions.

Table 4.6 ranks the average response for each variable assessed using Likert scale scores. These scores also range between twos and threes highlighting that the general perception of domestic environment factors that influence technological development is neutral or below-average levels of satisfaction. Intellectual property protection and the quality of ICT services are scored decently indicating sufficient policies to protect private sector interest in acceptable services derived from digital infrastructure. The remaining variables are below average and firms have a particularly low opinion of the availability and quality of R&D institutions for technical collaboration and alternative financing institutions including venture capital. The low scorer for the latter can also be attributed to the fact that most manufacturing firms in the sector rely on traditional financial institutions and funding mechanisms. There is a consistently poor perception of research and development institutions and governmental support for innovation.

Quality of Government and Institutional Support

Respondents were asked whether certain government institutions and functions explicitly benefitted the firms' ability to adopt new technologies, such as Industry 4.0-related technologies and to become more globally competitive. These functions included government-funded science and technology support institutions, testing and quality evaluation facilities, market research and intelligence support, overseas market promotion support, export credit programmes, financial incentives and grants, small and medium enterprise support, and inter-firm collaboration schemes.

Table 4.6 Quality of Domestic Environment for Technological Development, 2020–2021

Domestic Environment for Technological Development	Mean Response
Intellectual property protection	3.2
Quality of ICT services	3.0
Scientific/skilled manpower	2.6
Government incentives for innovation	2.5
Local universities for technical and R&D collaboration	2.4
R&D institutions for technical collaboration	2.4
Availability of venture capital	2.4

Source: Authors' survey.

Table 4.7 lists the average response for each variable assessed using Likert scale scores. For this section investigating the perceptions of the quality of government and institutional support, there is an overall unsatisfactory assessment. The overall mean scores were relatively low as they are all below 3.0. In comparison, direct support through financial incentives and grants and particular institutions designed for science and technology support, as well as overseas market promotion outperform others. Although the SME Corp was started by the government to support firms upgrading activities, the mean score of 2.4 suggests that its efforts have not received a significant number of beneficiaries, especially SME support for reforming inter-firm collaboration schemes. Support for market research and intelligence has been the weakest with a mean score of 2.2.

Connectivity and Coordination

Finally, respondents described the value of the relationship of their firms with particular stakeholders. These were the R&D organisations, financial institutions, distributors, suppliers, customers and end-users, technical service providers, business service providers, inter-firm relationships within industry associations, and the strength of strategic alliances.

Table 4.8 ranks the average response for each variable assessed using Likert scale scores. Across the four sub-sections, this is the only one that assesses the value of certain relationships with stakeholders against others. In this ranking, firms interviewed consider their relationships with customers and end-users, their suppliers, additional service providers, and financial institutions as the most valuable. Following that, the less important stakeholders are found within strategic alliances or industry associations, down the value chain to distributors, and finally, R&D organisations.

The rankings of these stakeholders reflect the dynamics within the electronics manufacturing industry where the demands of the customers and the linkages with other firms that support operational function and value chain flow continue to dictate the choices made within production facilities. The mean scores also show that the R&D linkages with external organisations are still relatively

Table 4.7 Government and Institutional Support for Technology Adoption and Competitiveness, 2020–2021

Government and Institutional Support	Mean Response
Financial incentives or grants	2.7
Science and technology support institutions	2.5
Overseas market promotion	2.5
Testing and quality evaluation facilities	2.4
Export credit programs	2.4
SME support and Inter-firm collaboration schemes	2.4
Market research and intelligence	2.2

Source: Authors' survey.

Table 4.8 Relationships with Stakeholders, 2020–2021

Relationships with Stakeholders	Average Response
Customers/end users	4.2
Suppliers of material & components	4.2
Technical service providers	3.9
Business service providers	3.7
Financial services institutions	3.7
Strength of strategic alliances	3.6
Relationship between firms in industry associations	3.6
Distributors	3.4
R&D organisations	3.0

Source: Authors' survey.

strong with a mean score of 3.0. Considering the individualistic and highly competitive nature of the sector, the mean scores show fairly strong links through strategic alliances and industry associations. However, the R&D linkages appear more to be functioning on an *ad hoc* basis with particular scientists at Malaysian universities.

Conclusions

In conclusion, Industry 4.0 is still at a nascent stage among Malaysian E&E firms. The ecosystem drivers that have the ability to encourage the adoption of these related technologies have been assessed and it shows that there are still significant weaknesses within the existing infrastructure, organisations, and policies. In particular, more needs to be done to enhance the quality of the research and development environment in Malaysia, though interactions between supporting organisations and firms are fairly strong. At the moment, the poor perception of the quality R&D institutions and support is hindering the potential for manufacturers in Malaysia to improve production capabilities and overall competitiveness. Additionally, this has previously and will continue to discourage investors from considering the nation as a potential destination for higher value-added production and product technologies in the highly competitive E&E industry. Universities and technical training institutes have much room for quality improvement in order to increase relevance in teaching and to become more active contributors to innovation within the domestic ecosystem.

This chapter offers several implications for policy to help enhance the ecosystem for the diffusion of Industry 4.0 technologies in the manufacturing sector. First, the Malaysian government in particular, and governments of developing countries in general, should solidify access to basic infrastructure while strategising to increase the accessibility and stability of modern digital infrastructure systems. Second, states can support high performing and emerging domestic firms that can provide related technical services for the development of such infrastructure and to other firms. Third, governments should focus on producing more industry-relevant content for public education and training institutes through collaboration with the private sector, regular syllabus review, and sufficient funding. Doing so fosters synergistic relationships and produces opportunities for quadru-lateral cooperation between government, industry, universities and institutes, and the labour force (see also Rasiah, 2019). Fourth, states can learn from innovative and compatible case studies based on other countries, such as Taiwan, for human capital development and modify these structures in accordance with domestic circumstances. The vocational training and reskilling and upskilling programmes in the country should review such successful models abroad.

Fifth, the Malaysian government in particular and developing country governments in general can continue to provide targeted financial and other policy incentives for research and development activity to deserving firms and institutions but by imposing stringent performance standards. Sixth, specific existing government linked support institutions should be evaluated for their relevance before conducting private sector consultations to target specific improvements in their services provided. This would be helpful for firms in the form of market research and intelligence, testing and quality evaluation, and overseas market promotion. Seventh, states can help to build firm synergy through organising more initiatives for inter-firm collaboration, especially for small domestic firms to find ways to become collectively more competitive. Finally, governments can help to promote these newer technologies by finding opportunities to link up the relevant advanced technology suppliers with manufacturers to raise awareness and allow a wider spread of knowledge. The focus should be on getting as much of firms involved rather than only focusing on those firms that are ready. Finally, the topic of Industry 4.0 should be promoted more widely amongst the general population because a more informed labour force and customer base will be both ready and also demand higher-quality products linked to better production processes.

Notes

1 The interviews were conducted by Yeap Khai Leang under the supervision and funding through the Distinguished Professor grant (MO-004-2018) held by Rajah Rasiah.
2 This was also to meet the ethics approval requirements.

References

Anderln, R. (2015) Industrie 4.0 – technological approaches, use cases, and implementation. *AT- Automatisierungstechnik*, Vol 63, Issue 10, pp. 753–765.

Armbrust, M., Fox, A., Griffith, R., Joseph, A.D., Katz, R., Konwinski, A., Lee, G., Patterson, D., Rabkin, A., Stoica, I., and Zaharia, M. (2010) A view of cloud computing. *Communications of the ACM*, Vol 53, Issue 4, pp. 50–58.

Ashton, K. (2009) That 'internet of things' thing, in the real world things matter more than ideas. *RFID Journal*, downloaded on February 5, 2022 from http://www.rfidjournal.com/articles/view?4986

Babiceanu, R.F., and Seker, R. (2016) Big Data and virtualization for manufacturing cyberphysical systems: A survey of the current status and future outlook. *Computer Industry*, Vol 81, pp. 128–137.

Bauer, W., Schlund, S., Hornung, T., and Schuler, S. (2018) Digitalization of industrial value chains – a review and evaluation of existing use cases of industry 4.0 in Germany. *LogForum*, Vol 14, Issue 3, pp. 331–340.

Benitez, G.B., Ayala, N.F., and Frank, A.G. (2020) Industry 4.0 innovation ecosystems: An evolutionary perspective on value cocreation. *International Journal of Production Economics*, 228, 107735.

Berry, H., Guillén, M.F., and Zhou, N. (2010) An institutional approach to cross national distance. *Journal of International Business Studies*, Vol 41, Issue 9, pp. 1460–1480

Brettel, M., Friederichsen, N., Keller, M., and Rosenberg, M. (2014) How virtualization, decentralization and network building change the manufacturing landscape. *International Journal of Mechanical Industry and Science Engineering*. Vol 8, pp. 37–44.

Brödner, P. (2018) Super-intelligent machine: Technological exuberance or the road to subjection. *AI and Society*, Vol 33, Issue 3, pp. 335–346.

Calabrese, A., Ghiron, N.L., and Tiburzi, L. (2020) 'Evolutions' and 'revolutions' in manufacturers' implementation of industry 4.0: A literature review, a multiple case study, and a conceptual framework. *Production Planning & Control*, Vol 32, Issue 3, pp. 213–227.

Castelo-Branco, I., Cruz-Jesus, F., and Oliveira, T. (2019) Assessing industry 4.0 readiness in manufacturing: Evidence for the European Union. *Computers in Industry*, Vol 107, pp. 22–32.

Chen, M., Wan, J., Gonzalez, S., Liao, X., and Leung, V.C.M. (2014) A survey of recent developments in home M2M networks. *IEEE Communications Surveys and Tutorials*, Vol 16, Issue 1, pp. 98–114.

Chiarello, F., Trivelli, L., Bonaccorsi, A., and Fantoni, G. (2018) Extracting and mapping industry 4.0 technologies using Wikipedia. *Computer Industry*, Vol 100, pp. 244–257.

Chiarini, A., Belvedere, V., and Grando, A. (2020) Industry 4.0 strategies and technological developments. An exploratory research from Italian manufacturing companies. *Production Planning and Control*, Vol 31, Issue 16, pp. 1385–1398.

Chryssolouris, G., Mavrikios, D., Papakostas, N., Mourtzis, D., Michalos, G., and Georgoulias, K. (2009) Digital manufacturing: History, perspectives, and outlook. *Proceedings of the Institution of Mechanical Engineering, Part B: Journal of Engineering Manufacture*, Vol 223, Issue 5, pp. 451–462.

Coase, R.H. (1937). The nature of the firm. *Economica*, Vol 4, Issue 16, pp. 386–405.

Cooper, J., and James, A. (2009) Challenges for database management in the Internet of things. *IETE Technical Review*, Vol 26, pp. 320–329.

Culot, G., Nassimbeni, G., Orzes, G., and Sartor, M. (2020) Behind the definition of industry 4.0: Analysis and open questions. *International Journal of Production Economics*, Vol 226, pp. 107617.

Dalenogare, L.S., Benitez, G.B., Ayala, N.F., and Frank, A.G. (2018) The expected contribution of industry 4.0 technologies for industrial performance. *International Journal of Production Economics*, Vol 204, pp. 383–394.

Damanpour, F. (1991) Organizational innovation: A meta-analysis of effects of determinants and moderators. *Academic Management Journal*, Vol 34, Issue 3, pp. 555–590.

Foidl, H., and Felderer, M. (2016) Research challenges of industry 4.0 for quality management. *International Conference on Enterprise Resource Planning Systems*, downloaded on April 2022, from https://link.springer.com/chapter/10.1007/978-3-319-32799-0_10

Frank, G., Dalenogare, L.S., and Ayala, F. (2019) Industry 4.0 technologies: Implementation patterns in manufacturing companies. *International Journal of Production Economics*, Vol 210, pp. 15–26.

Gibson, I., Rosen, D.W., and Stucker, B. (2014) Additive manufacturing technologies. *Springer*, Vol. 17, New York.

Gilchrist, A. (2016) Industry 4.0: The industrial internet of things. *Apress,* New York.

Hamilton, A., 1791. *Report on the Subject of Manufactures.* William Brown, Philadelphia, PA.

Hänninen, M., Smedlund, A., and Mitronen, L. (2018) Digitalization in retailing: Multi-sided platforms as drivers of industry transformation. *Baltic Journal of Management*, Vol 13, Issue 2, pp. 152–168.

Hermann, M., Pentek, T., and Otto, B. (2016) Design principles for industrie 4.0 scenarios. In *Proceedings of the 49th Hawaii International Conference on IEEE System Sciences (HICSS)*, Koloa, HI, USA, 5–8 January 2016.

Kagermann, H., Wahlster, W., and Helbig, J. (2013) Securing the Future of German Manufacturing: Recommendations for implementing the strategic initiative Industrie 4.0, Final report of the Industrie 4.0 Working Group, downloaded on April 3, 2022 from https://www.din.de/blob/76902/e8cac883f42bf28536e7e8165993f1fd/recommendations-for-implementing-industry-4-0-data.pdf

Lasi, H., Fettke, P., Kemper, H.G., Feld, T., and Hoffmann, M. (2014) Industry 4.0. *Business Information Systems Engineering*, Vol 6, pp. 239–242.

Laureti, T., Piccarozzi, M., and Aquilani, B. (2018) The effects of historical satisfaction, provided services characteristics and website dimensions on encounter overall satisfaction: A travel industry case study. *The TQM Journal*, Vol 30, pp. 197–216.

Lee, I., and Lee, K. (2015) The internet of things (IoT): Applications, investments, and challenges for enterprises. *Business Horizons*, Vol 58, Issue 4, pp. 431–440.

Lee, J., Bagheri, B., and Kao, H.A. (2015) A Cyber-Physical Systems architecture for industry 4.0-based manufacturing systems. *Manufacturing Letters*, Vol 3, pp. 18–23.

List, F. (1885) *The National System of Political Economy.* Longmans, Green & Company, London.

Moeuf, A., Lamouri, S., Pellerin, R., Tamayo-Giraldo, S., Tobon-Valencia, E., and Eburdy, R. (2019) Identification of critical success factors, risks and opportunities of industry 4.0 in SMEs. *International Journal of Production Research*, Vol 58, Issue 5, pp.1384–1400.

Mosconi, F. (2015) *The New European Industrial Policy: Global Competitiveness and the Manufacturing Renaissance.* Routledge, London, England.

Nelson, R. & Winter, S. (1982*) An Evolutionary Theory of Economic Change.* Belknap Press, Cambridge.

North, D. (1990) *Institutions, Institutional Change and Economic Performance.* Cambridge University Press, Cambridge, UK and NY.

Oesterreich, T.D., and Teuteberg, F. (2016) Understanding the implications of digitisation and automation in the context of Industry 4.0: A triangulation approach and elements of a research agenda for the construction industry. *Computer Industries*, Vol 83, pp. 121–139.

Pereira, A., and Romero, F. (2017) A review of the meanings and the implications of the Industry 4.0 concept. *Procedia Manufacturing*, Vol 13, pp. 1206–1214.

Pessot, E., Zabgiacomi, A., Battistella, C., and Rocchi, V. (2020) What matters in implementing the factory of the future: Insights from a survey in European manufacturing regions. *Journal of Manufacturing Technology Management*, Vol 3, Issue 3, pp. 795–819.

Rasiah, R. (2019) Building networks to harness innovation synergies: Towards an open systems approach to sustainable development, *Journal of Open Innovation: Technology, Markets and Complexity*, Vol 5, pp. 70.

Rasiah, R. (2020) Industry Policy and Industrialization in Southeast Asia, Arkebe, O., Cramer, C., Chang, H.J. & Kozul-Wright, R. (eds), *Oxford Handbook of Industrial Policy*, Oxford University Press, Oxford.

Rasiah, R., Rafat, B., and Yap, X.S. (2016) Institutional support, innovation capabilities and exports: Evidence from the semiconductor industry in Taiwan. *Technological Forecasting and Social Change*, Vol 109, Issue 8, pp. 69–75.

Rasiah, R., and Yap, X.S. (2016a) Institutional support, technological capabilities and domestic linkages in the semiconductor industry in Singapore. *Asia Pacific Business Review*, Vol 22, Issue 1, pp. 180–192.

Rasiah R., and Yap, X.S. (2016b) Institutional support, regional trade linkages and technological capabilities in the semiconductor industry in Malaysia. *Asia Pacific Business Review*, Vol 22, Issue 1, pp. 165–179.

Rauch, E., Unterhofer, M., Rojas, R.A., Gualtieri, L., Woschank, M., and Matt, D.T. (2020) Maturity level-based assessment tool to enhance the implementation of industry 4.0 in small and medium-sized enterprises. *Sustainability*, Vol 12, pp. 3559.

Roblek, V., Meško, M., and Krapež, A. (2016) A complex view of industry 4.0. *Sage Open*, Vol 6, Issue 2, pp. 1–11.

Rojko, A. (2017) Special focus paper – industry 4.0 concept: Background and overview. *International Journal of Interactive Mobile Technologies*, Vol 11, Issue 5), pp. 77–90.

Santos, M.Y., e Sá, J.O., Costa, C., Galvão, J., Andrade, C., Martinho, B., Lima, F.V., Costa, E., et al. (2017) A big data analytics architecture for industry 4.0. *World Conference on Information Systems and Technologies*, pp.175–184.

Schuh, G., Anderl, R., Gausemeier, J., ten Hompel, M., and Wahlster, W. (2017) Industrie 4.0 Maturity Index, Schuh, G., Anderl, R., Gausemeier, J., Hompel, M.T. and Wahlster, W. (eds), *Managing the Digital Transformation of Companies – Acatech Study*, Munich: Herbert Utz Verlag.

Smith, A. (1776) *An Inquiry into the Nature and Causes of the Wealth of Nations*. Random House, New York.

Stock, T., Obenaus, M., Kunz, S., and Kohl, H., (2018) Industry 4.0 as enabler for a sustainable development: A qualitative assessment of its ecological and social potential. *Process Safety and Environmental Protection*, Vol 118, pp. 254–267.

Sundberg, L., Gidlund, K.L., and Olsson, L. (2019) Towards Industry 4.0? Digital maturity of the manufacturing industry in a Swedish region. *Proceedings of the 2019 IEEE International Conference on Industrial Engineering and Engineering Management*.

Szalavetz, A. (2019) Industry 4.0 and capability development in manufacturing subsidiaries. *Technological Forecasting and Social Change*, Vol 145, pp. 384–395.

Vaidya, S., Ambad, P., and Bhosle, S. (2018) Industry 4.0 – A Glimpse. 2nd international conference on materials manufacturing and design engineering. *Procedia Manufacturing Edi*, Vol 20, pp. 233–238.

Veblen, T. (1915). *Imperial Germany and the Industrial Revolution*. Macmillan, New York.

Vrchota, J., Volek, T., and Novotná, M. (2019) Factors introducing industry 4.0 to SMEs. *Social Sciences*, Vol 8, pp. 130.

Wan, J., Tang, S., Li, D., Wang, S., Liu, C., and Abbas, H. (2017) A manufacturing big data solution for active preventive maintenance. *IEEE Transactions on Industrial Informatics*, Vol 13, Issue 4, pp. 2039–2047.

Wang, L., Törngren, M., and Onori, M., (2015) Current status and advancement of cyberphysical systems in manufacturing. *Journal of Manufacturing Systems*, Vol 37, pp. 517–527.

Wilkesmann, M. and Wilkesmann, U. (2018) Industry 4.0 – organizing routines or innovations? *VINE Journal of Information and Knowledge Management Systems*, Vol 48, Issue 2, pp.238–254.

Williamson, O.E. (1985) *The Economic Institution of Capitalism: Firms, Market, Relational Contracting.* Free Press, New York.

Yao, X., Zhou, J., Zhang, J., and Boer, C.R. (2017) From intelligent manufacturing to smart manufacturing for industry 4.0 driven by next generation artificial intelligence and further on. *Proceedings of the 2017 5th International Conference on Enterprise Systems: Industrial Digitalization by Enterprise Systems*, Article 8119409, pp. 311–318.

Zhong, R.Y., Xu, X., Klotz, E., and Newman, S.T. (2017) Intelligent manufacturing in the context of industry 4.0: A review. *Engineering*, Vol 3, pp.616–630.

5 Ecosystem Supporting Industry 4.0 Technologies in Textile and Clothing Manufacturing

Kiranjeet Kaur, Hemant Kedia, and Rajah Rasiah

Introduction

Schumpeter (1942) had argued that innovation is the driver of economic change as it generates gales of "creative destruction" which is often also the outcome of competition. Epochal changes have been the basis of broader innovation effects that support the substitution of one technological regime with another (e.g., the way railroads replaced canals and the trucking industry later challenging railroads). Indeed, Schumpeter (1942) glorified the large firm only because such firms enjoy the capital to afford the scale required to internalize R&D activities. Meanwhile, Perez and Soete (1988) proposed that the emerging technological paradigms should serve as a "window of opportunity" for latecomers to catch up. Consequently, one can view IR4.0 to present that "window" for countries or firms to seize opportunities previously thought impossible. The ramifications it would bring for manufacturing firms are to force governments to re-evaluate as to how the manufacturing industry operates and contributes to economic growth.

The world has experienced four industrial revolutions by 2022. The First Industrial Revolution (IR) (Figure 5.1) came about in the year 1760 and lasted until 1840, which was defined by the emergence of mechanization, steam power, and the introduction of shuttle water-jet weaving looms. The Second Industrial Revolution, (which lasted from the end of the 19th century to the first two decades of the 20th century), brought major breakthroughs in the form of electricity distribution, both wireless and wired communication, the synthesis of ammonia and new forms of power generation. The Third Industrial Revolution began in the 1950s with the development of automation, electronics, communication, and rapid advances in computing power, which have enabled new ways of generating, processing, and sharing information (Taalbi, 2018).

The Fourth Industrial Revolution (IR4.0) or simply Industry 4.0 saw the shift from the Third Industrial Revolution to the introduction of artificial intelligence in machines through the smartification of autonomous unmanned systems fuelled by data and machine learning (See Figure 5.1; Rasiah, 2019). The IR4.0 concept arose from the Hannover Fair in 2011, from when the German government officially announced it in 2013 as a German strategic initiative to

DOI: 10.4324/9781003367093-5

play a pioneering role in its manufacturing sector (Vaidya, Ambad & Bhosle, 2018), initially known as the "High-Tech Strategy 2020."

In the United Kingdom, the focus has been on strategic management of the process of IR4.0 formation and its role in the transformation of the market using such instruments, which is expected to make massive changes to consumer expectations and preferences under the influence of digitalization and IR4.0 technologies (Lobova et al., 2019). Meanwhile in Japan, the IR4.0 strategy is oriented towards gaining advantage related to the optimization of social systems, business-processes, and production of technologies and equipment, all of which are primarily to be driven by private business. The Japanese government's national strategy encourages businesses to develop and adopt long-term development strategies over five to ten years (Lobova et al., 2019).

The key components of IR4.0 is that it is surrounded by a vast network of advanced technologies across the value-chain, i.e., artificial intelligence (AI), robots, drones, internet of things (IoT), big data analytics, cloud computing, and additive manufacturing. The boundaries between the real world and virtual reality are getting blurrier, which has caused a phenomenon known as Cyber-Physical Production Systems (CPPS) (Schumacher, Erol & Sihna, 2016).

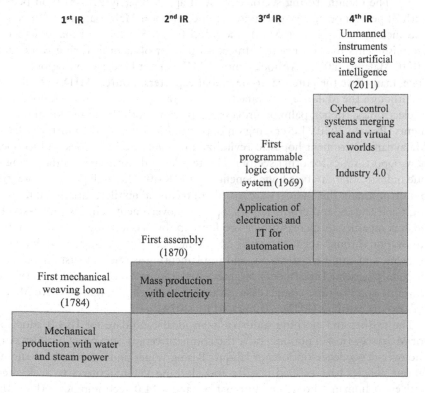

Figure 5.1 Industrial Revolutions and Key Themes.
Source: Adapted from Deloitte (2017: 2)

Industry 4.0 technologies have the potential to create extraordinary growth opportunities and competitive advantage for economic agents deploying them. Experts forecast that businesses will be able to increase their productivity by about 30 percent annually using Industry 4.0 technologies (EPICOR, 2020). Industry 4.0 builds upon several advanced technologies that depend on data generated by digitization of things, machines, and processes.

The National Policy on Industry 4.0 was launched on 31 October 2018 by the government to drive digital transformation of the manufacturing and related services sectors in Malaysia. The overarching philosophy behind this Policy is A-C-T – Attract, Create and Transform, i.e. attract stakeholders to Industry 4.0 technologies and processes; create the right ecosystem for Industry 4.0 technologies to be adopted and to nurture innovations; and transform capabilities of the manufacturing industry to be Industry 4.0-ready (MITI, 2020).

The manufacturing sector remains vital to most economies around the globe. For Malaysia, its contributions to the nation's export revenue and job creation helped the country's growth despite global economic uncertainties. Manufacturing is the second-largest contributor, after services, to Malaysia's GDP with over 49,000 establishments, and accounting for over 22 percent of the GDP in 2020. The manufacturing sector employed approximately 2.12 million people with 74 percent being Malaysians and 26 percent non-Malaysians in 2020. Small and medium enterprises (SMEs) accounted for 98.5 percent of manufacturing establishments and contributed almost 43 percent of manufacturing workers in 2020 (Malaysia, 2020). Although most SMEs do not have a strong global presence, many have the potential to be global exporters (Source: MIDA, 2020).

Although the Malaysian manufacturing sector is diverse it has been dominant in electronics, rubber, palm oil processing, pharmaceuticals, and medical instruments (Rasiah, 2020). Leveraging on opportunities offered by Industry 4.0, the Malaysian government hopes to revitalize many mature industries and to open new opportunities for other sectors. The textiles and clothing manufacturing is one industry that can be a huge beneficiary of IR4.0. The high growth potential opportunities in this industry are mainly in technical textiles, functional fabrics, high-end fabrics, and ethnic fabrics. Specific government vehicles are tasked to push companies to increase their productivity by accelerating automation and innovation through the use of artificial intelligence, undertaking research, development, and commercialization, and implementing green and sustainable production practices. Investments targeted at the improving product quality and lowering costs to raise competitive advantage are highly sought to drive Malaysia's export earnings (MIDA, 2020; MITI, 2022).

The textiles and clothing industry is an ideal candidate for IR4.0 adoption for Malaysia as it will not only raise the competitiveness of exports but also curb the over-dependence on foreign labour. Rising wages and frequently changing labour policies have pushed textiles and clothing manufacturers to seek alternatives to human labour, i.e. automation and IR4.0 technologies. Albeit the structure and processes have remained largely the same, textiles and clothing manufacturers have evolved into more technologically advanced versions of their predecessors by introducing computerized and programmable machines to replace humans. Some of the key technologies shaping the sector include 3D printing, body scanning technology, computer-aided design (CAD), wearable technology, nanotechnology, environmentally friendly manufacturing techniques, and robotization. Some parts of clothing manufacturing are already using sew-bots to replace low-skilled labour in tailoring. Also, advancements in nanoparticle research have introduced nanoparticle-infused clothes that are waterproof, stain-proof, UV protecting, and odourless. The impact of technology investment on productivity and growth is found to be greater at firm-level in comparison to industry and country-levels (Brynjolfsson & Hitt, 2003; Matteucci et al., 2005). At the firm level, technology use has led to improvements in product design, marketing, production, finance and the organization of firms, and development of new product (Becchetti et al., 2003; Carlsson, 2004; Hollenstein, 2004).

Malaysia has five designated economic corridor – Iskandar Malaysia (IM), East Coast Economic Region (ECER), Northern Corridor Economic Region, Sabah Development Corridor (SDC) and Sarawak Corridor of Renewable Energy (Score). Except for ECER and NCER, the other economic corridors are limited to the respective states only. IM was originally created as a gateway between Kuala Lumpur and Singapore for establishing a manufacturing and trading hub however missed the desired objective (Hutchinson, 2015; Rasiah & Gopi Krishnan, 2020) and it has turned more into real estate zone while the manufacturers are located outside of the economic zone (Rizzo & Glasson, 2012).

Textiles and Clothing Manufacturing

The textiles and clothing industry was a dominant industry in terms of employment, value of output and investment during the First Industrial Revolution, which began in Great Britain. In fact, the textile spinning machine invented by James Hargreaves in 1764 was one of the innovations that started the industrial revolution. The textile industry was also the first to use modern production methods (David, 1969) with mechanized factories, powered by central water wheel or steam engine. The Second Industrial Revolution was steam-powered and later the introduction of electrified manufacturing equipment led to the rise of textiles and clothing production in the late 19th century to the first truly industry-wide mass-production factories. During the Third Revolution in the 1960s up to 1980s, the introduction of electronic systems and computer technologies was another game changer for the industry. Auto-scanning, CAD, and computer-aided manufacturing helped lower throughput time and raise productivity.

The rapid diffusion of 3D printing, sensors, artificial intelligence, robotics, drones, and nanotechnology has continuously smartified manufacturing. While some of these technologies emerged before the turn of the millennium, the exponential expansion in computing power, reduction in cost, along with miniaturization has opened the way for the advent of big data analytics and cloud computing has made these technologies more suitable for industrial use (Deloitte, 2017). These are some of the technologies that has been proliferating in the textiles and clothing industry, which are capable of creating individual-ized pieces of clothing for consumers, while the production methods help raise production efficiency. Also, completely customized items produce zero waste materials, making it a lean operation by default. The increasing numbers and efficiency of smart factories support personalized clothing orders easy.

The textiles and clothing industry has been a significant contributor to the early growth of several economies as they have enjoyed rapid economic growth through connecting with the global value chains in the textiles and clothing industry, including in South Korea, Taiwan, Vietnam, Cambodia, Bangladesh, and China (Gereffi, 1999; Rasiah, 2012). Figure 5.2 shows the different com-ponents of the textiles and clothing value chain. Most least developed countries (LDCs) and developing countries are engaged in the manufacturing, design-ing, and logistics stages, while most brand holders are still in the developed countries.

The textiles and clothing industry has added value to each single segment in the supply chain beginning from the supplier of raw materials to the delivery of completed product to end consumer use (Khurana et al., 2010). It is intertwined with the agricultural sector for supply of raw materials i.e. natural fibres (such as cotton or wool), and the mining sector for the chemical fibres when it comes to the wide range of man-made fibres, such as nylon and polyester. Moreover, sev-eral other industries are consumers of technical textiles, which include products, such as filters, conveyer belts, optical fibres, packing textiles, ribbons and tapes, air bags, and insulation and roofing materials.

During the 1990s, the textiles and clothing industry was facing market mat-uration. This drove several firms to new low-cost locations in Asia that was also facilitated by the advent of internet technology, which improved coordination and control in remote poor locations (Djelic & Ainamo, 1999). This was further driven by the termination of the Multi-Fibre Arrangement in 2004 (Rasiah, 2012). However, the downside of this was the impact the move had on prod-uct quality and reliability, as well as its adverse impact on environmental and social conditions, such as poor working conditions, and unfair wages in the host countries (Robinson & Hsieh 2016), which drove the United States to impose the acceptance of the labour covenants of the International Labour Organiza-tion (ILO) and pressures on environmental and social conditions around the "everything but arms" clause by the European Union.

Given the substitution of humans with robots that has diffused into fron-tier firms, such as Tai Yuen and Everest Textiles, reshoring is occurring in some countries, such as Taiwan and the United States, but driven considerably by embedding science, technology and innovation (STI) infrastructure, and markets (see Nazia & Rasiah, forthcoming). However, given the extensive out-sourcing that has taken place and the nature of fiscal governance undertaken by individual governments, not all firms may relocate back to home-sites and major markets.

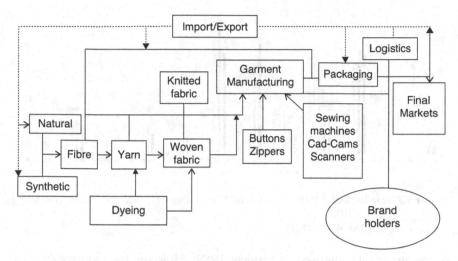

Figure 5.2 Textiles and Clothing Value Chain.
Source: Adapted from Rasiah (2012).

IR4.0 has presented new opportunities for businesses to rethink and redefine their business models. The technology it brings enables the creation of a new architecture of the entire ecosystem, where all data and information are collected and exchanged at any level of the organization and at any stage of the process throughout the entire value chain enabling the creation of a real-time virtual duplication of the whole system (Schwab, 2017). This paves the way for unprecedented benefits in terms of efficiency gains, agility, quicker turnaround time, consistency between customers' individual needs, and the product or service features. In short, IR4.0 is shifting production from the conventional manufacturing to smart factories, smart networks, and smart products.

Textiles and Clothing Exports

As can be seen in Figure 5.3, Malaysia's share of exports in world exports of textiles and clothing grew till 1990, but has largely plateaued since over the period 1990–2019. China remains the world's largest exporter, while exports from South Korea and Taiwan show a trend fall in the World's share of exports. Bangladesh and Cambodia have increased their share in world exports, primarily following the special provisions offered to these countries after the Multi-Fibre Agreement ended in 2004 (Rasiah, 2012).

Labour-intensive manufacturing has often been considered as the springboard of early industrialization in many developing countries which is why the textiles and clothing industry had a strong start in many Asian countries since the early 1960s. It played a key role in South Korea's early industrialization before other industries, such as automobiles, ships, electronics, and steel grew rapidly

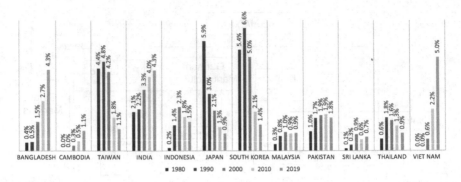

Figure 5.3 Textiles and Clothing Share in World Exports, Selected Asian countries, 1980–2019.

Source: Compiled from WTO (2021)

to overshadow its importance (Amsden, 1989). However, the industry remains crucial, and it covers a wide range of vertically integrated activities as South Korea remains a major exporter of textiles following its development of an extensive synthetic fibre industry.

Similarly, Taiwan had relied heavily on the textiles and clothing industry to drive its economy before machines tools, electronics, and medical instruments evolved to spearhead exports from the country (Wade, 1990). Taiwan began losing its comparative advantage in international production in 1990s especially for the labour-intensive parts of production, as such segments moved to China (Lee-in, 2007). While traditional production moved to China, Taiwan continued focusing on high-tech products. Through the Taiwan Textile Research Institute (TTRI), the government played a major role in technology-related services to transform the industry, such as technical guidance and talent training, testing and certification services, and industrial services. While polyester fibre was still a key product, leading firms started to develop polyester filament and nylon staple for industrial use with the R&D focused on producing hyper-thin fibres. Taiwan's core had shifted to artificial fibres and related weaving products. Some firms, such as Everest Textiles, managed to retain operations in the clothing industry through vertically integrated operations (Nazia & Rasiah, forthcoming). Taiwan has since become a leading synthetic fibre producer specialized in capital- and technology-intensive operations (Lee-in, 2007).

Vietnam's exports has grown commendably since its transition (*doi moi*) began in 1986. It has caught up and leap frogged Malaysia to record around 5 percent (USD 40.0 billion) of global exports compared to Malaysia's total exports of USD 7.2 billion in 2019. Textile and clothing has become the second-largest export earner for Vietnam over the period 2002–2022. The expansion in exports was particularly dramatic when Vietnam gained accession to the World Trade Organization (WTO) in 2006, despite the country having to compete without LDC trade privileges. Unlike many other developing countries, national capital owned around 60 percent of all private textile and garment firms (FES, 2017).

Textile and clothing sector in Malaysia was initially set up through foreign investment, especially from Japan, Hong Kong, and Taiwan. The first generation of textile entrepreneurs mostly were graduates from Taiwan's textile institutes. The Malaysian textiles and clothing manufacturing industry enjoyed a better start than most ASEAN countries, except Thailand. The industry peaked in around 1990 and has since plateaued with foreign labour playing an important role in sustaining exports since the 1990s. While textiles and clothing enjoys high potential for automation, firms in Malaysia have preferred foreign labour owing to low labour costs over automation. Nevertheless, the industry recorded around 98,000 employees in 2018 (MIDA, 2020). Also, the government still included textiles among the high-potential sectors under the Eleventh Malaysia Plan (See Figure 5.4), though its trade performance has continued to be modest, accounting for around 1 percent of total world exports over the period 1990–2019 (Figure 5.3).

The inclusion of textiles among the targeted industries by MIDA offer the industry the opportunity for accessing incentives, which it should seek through a shift towards IR4.0 technologies. Indeed, the proliferation of IR4.0 technologies in the world's frontier firms such as Tai Yuen, TAL Group, Toray Industries, and Everest Textiles shows that the is a strong future for the industry in Malaysia. The adoption of IR4.0 technologies in the industry shall also shed the dirty, dangerous, and demeaning categorization that the industry has faced before, besides reducing its reliance on low skilled foreign labour.

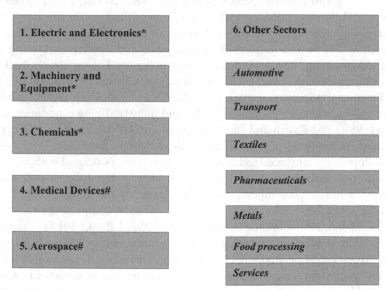

Figure 5.4 Focus Sectors, IR4.0 Promotion, Malaysia, 2020.
Note: Under the Eleventh Malaysia Plan, these sectors were identified as:
– 3 catalytic subsectors*
– 2 high potential growth sub-sectors#
Source: Adapted from MITI (2022), https://www.miti.gov.my/miti/resources/Infographic/Focus_Sector_on_Industry_4.0_.pdf

Theoretical Considerations

Three key models have been used in the past to evaluate the embedding eco-system supporting firms' activities, *viz.*, the diamond model (Porter, 1990), the productivity triad (Best, 2001), and the systemic quad (Rasiah, 2007). This chapter has preferred the systemic quad owing to its dynamics for assessing loose and little integrated clusters, and to focus on strengthening integration the instruments and organizations critical for stimulating upgrading in firms and for firms to enjoy systemic cluster synergies.

The Systemic Quad: A Model for Technology Upgrading

The systemic quad by Rasiah (2007, 2019) takes the centre stage in this chapter. It functions as a development model for a regionally networked group of economic agents, illustrating the process of building systemic support and their influence on knowledge related activities, advancement of technological capability, productivity as well as wages. The model comprises of four pillars which are: (1) basic infrastructure; (2) STI infrastructure; (3) network cohesion; and (4) global integration. It facilitates the evaluation of existing ecosystem and outlines distinctly the means by which underdeveloped ecosystems could be upgraded to support firm-level industrial upgrading.

Many theoretical studies emphasize that institutions are the fundamental source of long-run growth (Nelson & Winter, 1982; North, 1990; Acemoglu, Johnson & Robinson, 2005), which the function of the government is to ensure the integrated operations of various elements of industry policy (Armstrong & Sappington, 2006) to achieve their desired goals and objectives. During economic development, the government's role progressed from simply building basic infrastructure to creating STI infrastructure. Chang (1994) suggested that government participation in market regulation and supplying public goods to society and individuals were crucial in all rapid industrializing economies, while Rasiah (1995) and Rasiah and Lin (2005) had argued that trust and loyalty are important compliments of competition to drive firm performance.

The adoption of enhanced technologies is commonly associated with improved economic performance and development. In view of the growth potential of innovations bring, governments and many international development bodies across the globe have included the promotion of firms' competitiveness and the adoption of modern technologies as a priority (World Bank, 2017). Despite its potential, the process of technology adoption often can be sluggish (Geroski, 2000), which could be attributed by the high cost involved, the lack of knowledge and manpower, and risks and uncertainties involved to embrace new technologies. Hence, governments play an important role to assist the private sector to underwrite the risks and to share the public costs involved to support conducive ecosystems to catalyse firms' activities. Figure 5.5 shows the four systemic pillars that governments should focus on to stimulate technological upgrading in firms, including in quickening the diffusion of IR4.0 technologies.

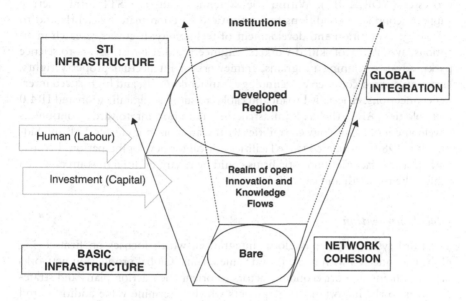

Figure 5.5 The Systemic Quad.
Source: Rasiah (2007).

Basic Infrastructure

The first pillar of basic infrastructure includes dynamic participation of government at federal, state, and local levels to provide macroeconomic and political stability, ensure security, transportation, communication networks, water, energy supply, health, education, and human capital. With efficient basic infrastructure, firms can focus on performing their operations in an organized way, which helps connect supply chains to efficiently move *goods* and services across borders. Aschauer (1989) had provided evidence to suggest that a reduction of public investment generally led to a fall in productivity growth. Efficient infrastructure strengthens economic growth and enhances quality of life, which is important for national security (Baldwin & Dixon, 2008). In addition, Levy and Spiller (1996) had demonstrated that superior institutional framework, i.e., independent regulatory authority, lower corruption, and contract enforcement, leads to better performance of economies.

Science, Technology, and Innovation Infrastructure

Rasiah (2007, 2019) had argued for the need to distinguish the STI infrastructure from basic infrastructure as the first enjoys public goods properties while the latter largely public goods utilities. Investments in infrastructure increasingly addresses the integration of technology in every aspect of the economy. Sophisticated public goods infrastructure is a prerequisite for innovative production processes (ADB, 2017). Within the systemic quad, the STI infrastructure supports conducive conditions for institutions to coordinate efficiently and to encourage acquisition and development of technology. It covers research institutions, availability of skilled workers, engineers and scientist, access to science parks, advance training programs, framework of intellectual property rights, laws regulating e-commerce, technology transfer councils, and high-speed internet connections, especially broadband cables to support digitalization and IR4.0 technologies. Also, the STI infrastructure is important towards continuous development of technology capabilities within an economic cluster. Lall and Siddhartan (1982) considered these facilities critical for developing nations because without these facilities domestic firms could be reliant on foreign sources which inhibit their growth and capabilities.

Global Integration

The third systemic pillar is global integration, which focuses on firms being globally connected with markets and value chains. Global production networks or value chains support economic actors to orient their action plans and strategic goals to the important developments which determine value additions and upgrading (Pietrobelli & Rabellotti, 2006). Integration can lead to substantial economic gains as it allows countries to improve market efficiency, access knowledge from multinationals through foreign direct investment (FDI) and attract policies for cooperation between organizations and firms across borders. Governments can help firms integrate through market promotion programmes be they in physical form or through e-commerce, such as export financing programmes, and sound framework to support international trade that is backed by strong financial instrument.

Connectivity and coordination globally with human capital, R&D laboratories, intellectual property rights organizations, and final markets helps enhance both competition and cooperation between firms, and in these case, textiles and clothing firms in their global value chains, to support globally competitive operations. The internet, especially digital infrastructure, has been a key driver of global market coordination (Nazia & Rasiah, forthcoming).

Network Cohesion

The last pillar is network cohesion which outlines the importance of encouraging the interdependence and interaction between various parties involved in not just spatially defined clusters, but also with international buyers and suppliers, and organizations. Cohesive networks also facilitate access to information and knowledge that is not readily accessible in markets (Rasiah, 2007). It creates opportunities for acquiring and exploiting new knowledge required for innovation (Rasiah, 2019). In a business setting, firms network with different agents within a cluster or supply chain i.e. firm – R&D centres, financial institutions, suppliers of raw materials, distributors and sellers, service providers, trade associations as well as industry peers. In addition, trust and loyalty factors were recognized as factors that can further strengthen the network cohesion pillar (Rasiah & Lin, 2005).

A popular notion among researchers of clusters is that firms in developed geographic clusters have higher levels of performance because of superior support institutions, organizations, and cohesive interconnections between them and firms. Such integrated clusters help movement of purchases and sales, labour, and knowledge embodied in humans, machinery, products, capital, and processes smoothly between economic agents (Schoonhoven & Eisenhardt, 1990; Saxenian, 1990; Shaver & Flyer, 2000; Porter & Stern, 2001; Rasiah, 2007).

While the four systemic pillars are critical for assessing the ecosystem supporting textiles and clothing firms in Malaysia, it is also pertinent to note the importance of knowledge networks that has facilitated the small firms' participation in core innovation activities. Schumpeter (1942) had argued that superior large firm size and the market structure in which firms operate in is the main engine of technological progress. In contrast, Acs and Audretsch (1987, 1990) found that smaller firms tend to have higher innovation intensity compared to larger firms. As Rasiah (2019) and Rasiah and Nazia (forthcoming) had argued, Schumpeter did not envisage the development of knowledge networks that externalized the innovation process so that small firms could connect with R&D laboratories in universities, the military and other public bodies, as well as incubators at universities and science parks to appropriate innovation synergies. Meanwhile, Acs and Audretsch's (1990) claim that smaller firms rather than larger firms enjoy higher innovation intensity is not founded on the use of a rigorous methodology that controlled effectively for industry-type and the composition of the embedding ecosystem supporting innovation activities. Consequently, this chapter focuses on the embedding ecosystem supporting the shift to IR4.0 technologies in textiles and clothing firms in Malaysia. Digitization is the prerequisite of Industry 4.0.

Methodology and Data

Drawing on the systemic quad, this chapter assesses the performance of key enablers that can facilitate Malaysia's efforts to capitalize on emerging technologies and opportunities presented by it. Table 5.1 gives a breakdown of the four key pillars and the enablers that make up these pillars. The prime focus of the study is to evaluate the effectiveness of the embedding ecosystem to support the operations of textiles and clothing firms, especially to support absorption of IR4.0 technologies.

Data collection was done through face-to-face interviews and phone calls, as well as zoom interviews using probabilistic sampling technique. The lists of firms were obtained from two of the largest textile and clothing associations in Malaysia, *viz.*, Malaysian Textile Manufacturing Association (MTMA) and Malaysian Knitting Manufacturing Association (MKMA).

Only manufacturers from the upstream and downstream segments were selected from the 115 manufacturers in the joint list, though due to the

Table 5.1 Systemic Pillars, Ecosystem Support Components

Basic infrastructure	Water and electricity supply
	High-speed broadband internet availability
	Public health facilities access
	Coordination within local council
	Access to capital/ credit from government
	Quality of education
	Labour regulations
	Political Stability
	Roads & seaports
STI infrastructure	Government incentive for innovation
	Quality of engineers and scientist
	University / research institution collaboration
	Access to science parks
	Intellectual property protection laws
	Advanced training programs
	Regulations For E- commerce
	Access to technology transfer councils
Global integration	Overseas market promotion
	Online commerce platform to access global market
	Government backed export financing programs
	Sound legal framework for international trade
	Strong domestic banking infrastructure to support firms
	Availability of alternative finance mechanism i.e., crowd funding
Network cohesion	Firm – R&D organizations
	Firm – Financial institutions and banks
	Firm – Suppliers of materials and components
	Firm – Distributors and sellers
	Firm – Technical service provider
	Firm – Association
	Firm – Industry peers

Source: Adapted from Rasiah's (2004) UNU-INTECH Survey.

pandemic, many firms were closed. More than 80 firms were contacted via email and 52 responses were recorded which represents approximately 45 percent of the population among the MTMA and MKMA manufacturing members. The respondents consisted of the leading textiles and clothing firms in the country who make up more than 80 percent of the equity in the sector. Eventually, a total of 22 textile manufacturers and 30 clothing manufacturers participated in the study.[1] There were equal numbers of large firms and SMEs, i.e., 26 firms each, respectively. Firm-size in this study were based on Bank Negara Malaysia's classification, i.e., SME = 0–200 employees and Large firms = 201 and above employees.

There are 22 textile firms in the sample, and they are mostly made up of foreign firms engaged in yarn spinning, knitting, weaving, and fabric finishing. Only one large local firm was operating in this segment, indicating a huge gap in local capabilities in textile manufacturing. In contrast, clothing manufacturing is dominated by local firms, i.e., 22 out of the 30 firms interviewed are local firms. However, 78 percent of these local firms are SMEs and only 22 percent are large firms. Among foreign firms, 88 percent are large and there remaining

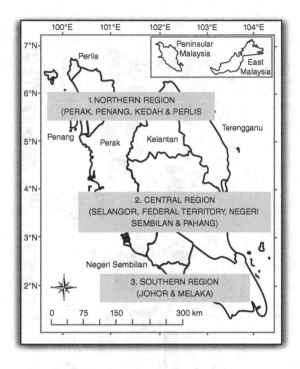

Figure 5.6 Location of Respondents, West Malaysia, 2020.
Source: Authors' survey.

Table 5.2 Respondents, Sample, 2020

Category	N	Size of Firms (as per BNM Definition)	Total	Breakdown by Ownership		Respondent's Profile
				Local	Foreign	
Textile (Yarn spinning, knitting / weaving / Fabric finishing) – includes home and technical textile players	22	Small & medium (up to 200 employees)	8	7	1	30% – Owners / C-Level executives
		Large (> 201 employees)	14	1	13	60% – Production managers
Clothing	30	Small & medium (up to 200 employees)	18	17	1	10% – Business Dev & strategy head
		Large (> 201 employees)	12	5	7	
Total	52					

Note: Size of firms is based on Bank Negara Malaysia's description i.e. SME = 0–200 employees and Large firms = 201 and above employees.
Source: Authors' Survey.

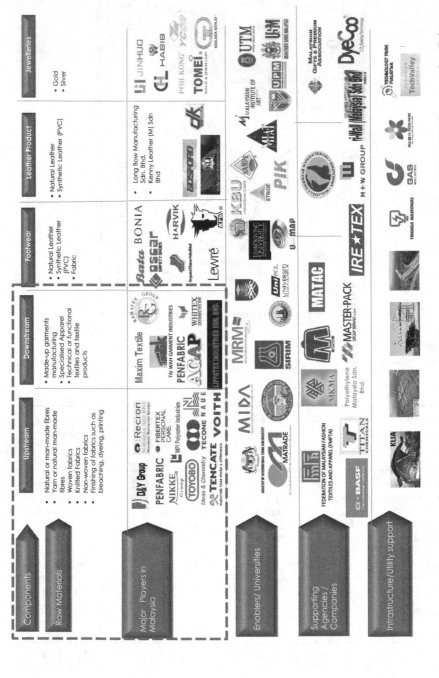

Figure 5.7 Ecosystem Supporting Textiles and Clothing Firms, Malaysia, 2022. Source: Adapted from MIDA (2022), accessed on February 9, 2022 from www.mida.gov.my.

12 percent are small or medium sized. The firms were in three regions across West Malaysia – northern, central and southern region as shown in Figure 5.6. As there was only one firm from the state of Pahang, we have grouped it central. By ownership, 30 firms are nationally owned and 22 are foreign-owned. All joint-venture between national and foreign capital were classified on the basis on majority equity ownership.

Likert-scale scores of 1 = Strongly disagree to 5 = Strongly agree were used to evaluate the strength of connectivity and coordination between firms and the organizations and instruments. For brevity, mean scores between 3.0 and 5.0 were classified as between good to excellent, while scores between below 3.0 to 1 were considered average to poor. Location and ownership are introduced to investigate how these elements impact the ratings of the ecosystem as there could be instances where certain clusters or location may experience superior infrastructure compared to the rest of the population. Clusters that perform well across the various enablers are considered more prepared to absorb IR4.0 technologies because the mix of enablers will allow for the adoption and diffusion of technology to accelerate transformation of the firm.

Ecosystem Supporting Textiles and Clothing Firms

This section discusses the appraisal of the four systemic pillars by the 52 textile and clothing players. While the textile and clothing sector is considered as one, it is important to note that there are varying needs and priorities of each segment. Hence, the paper will also assess these two sub-sectors. Another interesting finding from the research is that the textile industry is largely foreign-owned, while clothing is mostly locally owned. Hence, the segregation will allow us to understand how each segment rates the ecosystem and then on recommend possible policy interventions and improvements for the respective segments.

Basic Infrastructure

The overall rating of basic infrastructure facing textiles and clothing firms is generally good at the sub-sector level, ownership type as well as firm size. However, for assessment by location, the southern region scored between average to poor in all components of basic infrastructure. Also, the mean scores for capital/credit support enjoyed average to poor support in all horizontal categories. Nevertheless, except for the northern region on coordination councils, and the southern region in all horizontal categories, the remaining components of basic infrastructure were rated between good to excellent by the responding textiles and clothing firms (Tables 5.2 and 5.3).

As a case in point, it appears that the southern region requires significant improvement in basic infrastructure. One of the largest vertically integrated firm in Malaysia based in the southern region revealed that due to the poor state of infrastructure, it has resorted to building its own private 32 kilovolt (KV) substation and own roads leading to the factory equipped with traffic lights

Table 5.3 Basic Infrastructure, Textile and Clothing Firms, Malaysia, 2020

Basic Infrastructure	Assessment by Sub Sector		Assessment by Ownership		Assessment by Firm size		Assessment by Firm Location		
	Textile (N=22)	Clothing (N=30)	Foreign (N=12)	National (N=30)	Large (N=26)	SME (N=26)	Central (N=28)	South (N=16)	North (N=8)
Water/Electricity supply	3.86	3.90	3.77	3.97	3.31	3.92	3.89	2.94	4.00
High-speed internet	4.14	3.87	3.95	4.00	3.35	4.12	4.25	2.56	4.25
Public health facilities	3.18	3.43	3.41	3.27	2.92	3.38	3.25	2.75	3.63
Coordination within local council	3.32	2.93	3.09	3.10	2.65	3.08	3.11	2.31	3.13
Access to capital/credit from govt	2.32	2.53	1.77	2.93	1.62	2.88	2.36	1.88	2.63
Quality of Education	3.18	3.17	3.36	3.03	2.54	3.35	3.21	2.25	3.38
Labour regulations	2.91	2.37	2.41	2.73	2.23	2.69	2.57	2.13	2.75
Political stability	3.77	2.90	3.45	3.13	3.15	2.96	3.14	2.63	3.63
Road/Seaports	3.59	3.10	3.32	3.30	2.96	3.19	3.11	2.88	3.38

Source: Calculated from Authors' Survey.

to ensure safer travel around the vicinity. The company also pays for TNB's horizontal directional drilling (HDD). According to the spokesperson, despite being the largest employer in the area i.e. more than 12,000 people employed presently it enjoys no special privileges from the local government. Internet service is also rated poorly among firms in the Southern Region. However, both Northern and Central Region enjoyed better internet services. According to a firm located in the Batu Pahat industrial park, the current internet speed in the area is about 200 megabyte (MB). While there are efforts to upgrade the internet connectivity nationwide, the current state leaves much to be desired. Under the Malaysian Digital Blueprint (MDB), broadband is mandated as basic infrastructure to ensure internet access for all new developments (MDB, EPU, 2019).

Similarly, while there are numerous financing schemes by the government for national firms especially in the SME segment, many firms noted that the conditions set by banks and other financial institutions are too stringent so as to disqualify most of them for these schemes. Many must resort to commercial financing by banks that are collateral backed. One of the popular government schemes to help SMEs is the Credit Guarantee Corporations (CGC), through which the government guarantees loan applications by SMEs who are categorized as high risk. However, the margins charged by CGC in addition to the rates charged by commercial banks makes the scheme unattractive for SMEs.

Some of the respondents are unaware of the schemes that may be available to them and depend on informal information sources. While there are numerous workshops conducted by Government through partnerships with commercial banks, some firms admitted that they have never been invited textiles and clothing have not been considered as attractive by banks. The Risk Weighted Asset (RWA), which refers to an asset classification system that is used to determine the minimum capital that banks should keep as a reserve to reduce the risk of insolvency, is high for the textiles and clothing industry causing further strain in securing financing.

Science, Technology, and Innovation Infrastructure

STI infrastructure is critical to technological upgrading. Assessment by sub sectors rated the current state and availability of these enablers as weak. Clothing sector which is SME-dominated lacked government incentives for innovation. When reviewing assessment by location, Penang recorded the highest score compared to Central and South.

The mean Likert-scale scores of firms' ratings on the STI infrastructure was lower than that for basic infrastructure (Table 5.4). The northern region enjoyed the best ratings, though the scores for university-industry collaboration, access to science parks, and access to technology transfer councils, and advanced training programmes were rated below 3.0. Clothing firms enjoyed scores below 3.0 under all horizontal STI categories. Also, textiles firms enjoyed above 3.0 rating for access to innovation incentives, IP protection laws, and quality engineers and scientists.

IP protection laws is an important instrument that protects to promote innovation. Textiles firms rated this enabler high compared to clothing firms. Several textile players have patented their products unlike the clothing players where innovation is at firm level (Table 5.4). The assessment for regulation for e-commerce showed Penang Region scoring the highest. Overall infrastructure to attract FDI in this sector remains weak. Some foreign firms cited that they were looking at Vietnam and Bangladesh as an alternative location to set up operations as Malaysia does not provide much support to players in this sector.

Penang was rated highest for the quality of engineers and scientists. The island state's active involvement in the global value chain through the multinational corporations (MNCs) operating there has helped Penang nurture world class manpower (Rasiah, 1995; Lai & Narayanan, 1999; Athukorala, 2017). On the contrary, Johore firms face steep competition from Singapore in terms of hiring and keeping good talent as the latter offers significantly better remuneration and prospects. In the central region, good talent is spoilt for choice when it comes to employment as Selangor and Kuala Lumpur is home to many esteemed companies. Many firms cited that finding good talent is also a challenge especially in the clothing sector as it is not appealing to the younger generations, which explains why around 30 percent of workers are not only from abroad but are also low-skilled. Institutions offering degree programmes associated with textiles and clothing training are limited and few. A quick check revealed there are only two universities offering courses that caters for the textiles and clothing industry, i.e., University Technology Mara (UiTM) and Tun Hussein Onn University (THOU). However, due to the lack of student enrolment, UiTM is contemplating discontinuing that programme.

Collaboration with Universities or Research Institutions also is rated low across all aspects – by ownership, size, and location. There are however pockets of *ad hoc* collaboration happening between firms and universities. One such collaboration is between a bamboo fabric manufacturer with THOU to produce Malaysian grown bamboo that can be converted into fabric as the firm currently imports the raw materials from China. Also, access to the incubators at science parks and links with the technology transfer council is rated low as the textiles and clothing firms are not listed among the promoted activities. Technology Park Malaysia in Bukit Jalil is set as the country's innovation ecosystem one-stop centre and is entrusted to set up the Industrial Revolution 4.0 International Innovation Hub (IIH). TPM has been allocated RM30 million under Budget 2022 to establish the IIH, which would act as a centre of excellence (COE) and support the innovation ecosystem by developing new technology clusters such as drones, robotics, and autonomous vehicles (Star, 2021).

Global Integration

Global Integration is important in the textiles and clothing industry as it has connected their products, knowledge and technology between producers and buyers, including brand holders and final consumers (Future of Production Report,

Table 5.4 STI Infrastructure, Textile and Clothing Firms, Malaysia, 2020

Science, Technology, and Innovation Infrastructure	Assessment by Sub Sector		Assessment by Ownership		Assessment by Firm size		Assessment by Firm Location		
	Textile	Clothing	Foreign	National	Large	SME	Central	South	North
	(N=22)	(N=30)	(N=22)	(N=30)	(N=26)	(N=26)	(N=28)	(N=16)	(N=8)
Govt incentive for innovation	3.32	2.37	3.14	2.50	2.85	2.38	2.32	2.75	3.38
Quality engineers / scientist	3.32	2.67	3.14	2.80	2.81	2.77	2.61	2.69	3.63
University /research instcollab	2.68	2.17	2.50	2.30	2.15	2.27	2.14	2.31	2.25
Access to science parks	2.55	1.77	2.23	2.00	1.85	2.08	1.93	1.94	2.13
IP protection laws	3.45	2.80	3.32	2.90	3.00	2.96	2.82	2.81	3.88
Advanced training program	2.50	1.60	2.14	1.87	2.04	1.81	1.89	1.56	2.75
Regulations for ecommerce	2.86	2.97	2.77	3.03	2.69	2.96	2.75	2.81	3.13
Access to tech transfer councils	2.36	2.10	2.55	1.97	2.00	2.08	2.07	1.88	2.25

Source: Calculated from Authors' Survey.

2018). Trade liberalization has been the main driver behind the integration of developing countries in the global economy (Krueger, 1997), though states have played critical roles to stimulate industrial upgrading among national firms (Rasiah, 2020).

With the exception of the southern region, firms with links with domestic banking infrastructure, online platforms to access global markets, and legal framework for international trade scored either close 3.0 or more (Table 5.5). Unlike until the 1990s, government support for overseas market promotion, and government-backed financing and alternative financing for market promotion also scored between low to average.

The textiles and clothing industry in Malaysia is export-oriented with the majority of firms operating as original equipment manufacturers (OEMs). Since the United States is the export market IP protection laws have been an important instrument for promoting innovation. Textiles firms rated IP as a high-export enabler compared to clothing firms. Several textile players have patented their products. The assessment for regulation for e-commerce showed Penang Region scoring the highest. Overall, the infrastructure to attract FDI in this sector remains weak. Some foreign firms cited that they were looking at Vietnam and Bangladesh as an alternative location to set up operations as Malaysia does not provide much support to firms in the industry.

MIDA actively works at attracting foreign investments into the country. It did play a critical role until the 1990s to attract textiles and clothing firms to Malaysia. Although MIDA has set aside RM250 million in grants for the automation in the industry, textiles and clothing firms have not enjoyed accessed to those grants. Also, textiles and clothing firms enjoy little access to funds mobilized by the government for overseas market promotion. Nevertheless, the assessment by ownership yielded better results for national firms compared to foreign firms, including MATRADE's online export matching services for Malaysian exporters seeking foreign buyers.

Network Cohesion

Firms tend to cluster in a certain location to leverage on reduced transport costs, shared inputs, and productivity spill over from learning and technology transfer. Clustering can facilitate agglomeration economies, i.e., firm-level productivity gains that come from spatial concentration of economic activity (Krugman, 1991; Newman & Page, 2017). Textiles and clothing firms rated the relationship with financial institutions and banks, firm-supplier links for materials and components, and distributors and sellers, and firm-level technology providers as between close to average to excellent, regardless of sub-industry, ownership, firm-size, and location (Figure 5.7). However, firms reported below average links with R&D organizations and universities in all sub-categories. While the firms reported good to excellent their relationship with the industry associations by the industry sub-category of textiles and clothing, ownership, and size, only the central region enjoys good links in the location category (Table 5.6).

Table 5.5 Global Integration, Textile and Clothing Firms, Malaysia, 2020

Global Integration	Assessment by Sub Sector		Assessment by Ownership		Assessment by Firm Size		Assessment by Firm Location		
	Textile (N=22)	Clothing (N=31)	Foreign (N=22)	National (N=30)	Large (N=26)	SME (N=26)	Central (N=28)	South (N=16)	North (N=8)
Overseas market promotion	2.73	2.53	2.18	2.93	1.96	2.88	2.64	1.94	2.63
Online platform to access global market	2.95	3.23	2.95	3.23	2.73	3.15	2.89	2.88	3.25
Govt.-backed export financing programs	2.41	1.87	2.05	2.13	1.92	2.00	1.86	1.88	2.50
Sound legal framework for int. trade	3.45	3.00	3.32	3.10	2.96	3.08	3.14	2.63	3.38
Strong domestic banking infra	3.23	3.30	3.27	3.27	2.92	3.23	3.32	2.56	3.25
Availability of alternative finance	2.27	1.87	2.23	1.90	1.96	1.81	1.86	1.94	1.88

Source: Calculated from Authors' Survey.

Table 5.6 Network Cohesion, Textile and Clothing Firms, Malaysia, 2020

Network Cohesion	Assessment by Sub Sector		Assessment by Ownership		Assessment by Firm Size		Assessment by Firm Location		
	Textile	Clothing	Foreign	National	Large	SME	Central	South	North
	(N=22)	(N=30)	(N=22)	(N=30)	(N=26)	(N=26)	(N=28)	(N=16)	(N=8)
Firmrd orga/Uni	2.18	2.47	2.55	2.20	2.12	2.23	2.18	2.13	2.25
Firm with financial inst/banks	3.82	3.60	4.05	3.43	3.54	3.42	3.64	3.06	3.75
Firm suppliers of materials & components	4.00	4.40	4.32	4.17	4.00	4.00	4.14	3.50	4.50
Firm with distributors/sellers	4.41	4.27	4.68	4.07	4.04	4.15	4.29	3.44	4.75
Firm technical service provider	3.59	3.90	4.14	3.50	3.54	3.58	3.82	2.94	3.88
Firm association	3.55	3.37	3.68	3.27	3.12	3.38	3.61	2.81	2.88
Firm with peers	3.68	3.20	3.82	3.10	3.12	3.27	3.36	2.94	3.13

Source: Calculated from Authors' Survey.

Firms located in the central and northern regions show higher quality of connections and coordination with critical organizations because of superior basic and STI infrastructure embedding them. Textiles and clothing firms in the southern region are largely located around Muar and Segamat, which is among the least developed parts of Johore. Although the federal treatment of these firms is the same as elsewhere in Malaysia, there are no proximate universities or science and technology parks in the southern region. Consequently, the lack of connections between the firms and organizations, including among broader associations, such as the Federation of Malaysian Manufacturers (FMM), (which is headquartered in Kuala Lumpur), and the MKMA and MTMA. Nevertheless, proliferation of IR4.0 technologies, that is increasingly freeing the textiles and clothing industry from dirty, dangerous, and demeaning industries, offers the country to catch the wave of change to reinvigorate the industry to upgrade to the category of high value-added industries.

Conclusions

This chapter used the systemic quad to analyse the strength of the ecosystem supporting textiles and clothing manufacturing firms in Malaysia. Using the four systemic pillars, i.e., basic Infrastructure, STI infrastructure, global integration, and network cohesion, it measures how well-positioned are the textiles and clothing firms across various locations throughout Malaysia to absorb IR4.0 technologies in a country where few firms have seriously sought to mobilize resources to absorb IR4.0 technologies, which has been dampened by the outbreak of the COVID-19 pandemic (see also Rasiah, 2021). The firm-level appraisal shows that, except for the southern region, the basic infrastructure embedding the remaining locations is generally fine except for funding facing textiles and clothing firms.

Whereas the evidence shows that the basic infrastructure facing textiles and clothing firms is either good or excellent, except for the southern region, which scored between average to poor. Also, access to subsidized finance was rated poor among all horizontal categories. On the STI infrastructure pillar, collaboration with universities and research organizations was also rated low in all categories. Also, access to the incubators at science parks and links with the technology transfer council is rated low as the textiles and clothing firms are not listed among the promoted activities. Textiles and clothing firms rated the relationship with commercial financial organizations and banks, firm-supplier links for materials and components, and distributors and sellers, and firm-level technology providers as between close to average to excellent. While the firms reported good to excellent their relationship with the industry associations by the industry sub-category of textiles and clothing, ownership, and size, only the central region enjoys good links in the location category. Firms operating in the central and northern regions show higher quality of connections and coordination with critical organizations because of superior basic and STI infrastructure embedding them. The lack of STI organizations, such as training centres and universities, has reduced network cohesion between firms and critical intermediary organizations in Southern region.

The findings concur with the findings of Rasiah (2020a, 2020b), i.e., that the basic infrastructure and integration with global factor and final markets of firms in Malaysia are generally good but much more should be done to strengthen the STI infrastructure and network cohesion between the firms and universities and R&D organizations to strengthen the embedding ecosystem supporting the firms. Such an institutional upgrading should be undertaken simultaneously with the promotion of IR4.0 technologies in firms.

Note

1 The interviews were conducted by Kiranjeet Kaur under the supervision and funding through the Distinguished Professor grant (MO-004–2018) held by Rajah Rasiah.

References

ADB (2017). *Asian Development outlook*, Manila: Asian Development Bank.

Acemoglu, D., Johnson, S., & Robinson, J.A. (2005). The rise of Europe: Atlantic trade, institutional change and economic growth. *American Economic Review*, 95(3): 546–579.

Acs, Z.J., & Audretsch, D.B. (1987). Innovation, market structure, and firm size. *Review of Economics and Statistics*, 69(4): 567–574.

Acs, Z.J., & Audretsch, D.B. (1990). The determinants of small-firm growth in US manufacturing. *Applied Economics*, 22(2): 143–153.

Amsden, A.H. (1989). *Asia's Next Giant: South Korea and Late Industrialization*. New York: Oxford University Press.

Armstrong, M., & Sappington, D. (2006). Regulation, competition, and liberalization. *Journal of Economics Literature*, 44(2): 325–366.

Aschauer, D.A. (1989). Is public expenditure productive? *Journal of Monetary Economics*, 23(2): 177–200.

Athukorala, P.C. (2017). China's evolving role in global production networks: Implications for Trump's trade war. *China's New Sources of Economic Growth: Human Capital, Innovation and Technological Change*, 2: 363–388.

Baldwin, J.R., & Dixon, J. (2008). Infrastructure capital: What is it? Where is it? How much of it is there? *Canadian Productivity Review*. No 16. Ottawa: Statistics Canada.

Becchetti, L., & Sierra, J. (2003). Bankruptcy risk and productive efficiency in manufacturing firms. *Journal of Banking & Finance*, 27(11): 2099–2120.

Best, M.H. (2001). *The New Competitive Advantage: The Renewal of American Industry*, Oxford: Oxford University Press.

Brynjolfsson, E., & Hitt, L.M. (2003). Computing productivity: Firm-level evidence. *Review of Economics and Statistics*, 85(4), 793–808.

Carlsson, B. (2004). The digital economy: what is new and what is not?. *Structural Change and Economic Dynamics*, 15(3), 245–264.

Chang, H.J. (1994). *The Political Economy of Industrial Policy*. Basingstoke: Macmillan.

David S. Landes (1969). *The Unbound Prometheus*. Cambridge: Cambridge University Press.

Djelic, M.L., & Ainamo, A. (1999). The coevolution of new organizational forms in the fashion industry: A historical and comparative study of France, Italy, and the United States. *Organizational Science*, 10(5), 622–637.

Chang, H.J. (1994). *The Political Economy of Industrial Policy*. Basingstoke: Macmillan.

Deloitte (2017) *Birth of a Smart Factory*, Accessed on June 15 from https://www2. deloitte.com/content/dam/Deloitte/cy/Documents/innovation-and-entrepreneurship-%20centre/CY_IEC_Industry4.0_Noexp.pdf

Eisenhardt, K.M., & Schoonhoven, C.B. (1990). Organizational growth: Linking founding team, strategy, environment, and growth among US semiconductor ventures, 1978–1988. *Administrative Science Quarterly*, 35(3): 504–529.

EPICOR (2020). *What Is Industry 4.0?*. Accessed on June 14 from https://www.epicor. com/en-my/blog/categories/?IsFilter=Y&q=1091

FES (2017). The future of the Cambodian garment and footwear industry, July 20. *Phnom Penh: Fridrich-Ebert Stiftung (FES)*. Accessed on June 14 from https://asia.fes. de/news/the-future-of-the-cambodian-garment-and-footwear-industry

Gereffi, G. (1999). International trade and industrial upgrading in the apparel commodity chain. *Journal of International Economics*, 48(1): 37–70.

Geroski, P.A. (2000). Models of technology diffusion. *Research Policy*, 29(4–5): 603–625.

Hollenstein, H. (2004). Determinants of the adoption of information and communication technologies (ICT): An empirical analysis based on firm-level data for the Swiss business sector. *Structural Change and Economic Dynamics*, 15(3): 315–342.

https://www.mida.gov.my/industries/manufacturing/

Hutchinson, F.E (2015). *Mirror Images on Different frames? Johor, the Riau Islands and Competition for Investment in Singapore*. Singapore: Institute of Southeast Asian Studies (ISEAS).

Johnson, C.A. (1982). *Revolutionary Change Vol. 47*. Stanford: Stanford University Press.

Khurana, P., Pahwa, M., Bansal, A., Dhingra, R., & Sharma, A. (2010). *Indian Garment Industry*. Accessed on July 5, 2013 from http://www.scribd.com/doc/49342458/INDIAN-GARMENT-INDUSTRY.

Krueger, A.O. (1997). Trade policy and economic development: How we learn, National Bureau of Economic Research, Working Paper No. 5896, downloaded on April 15, 2022 from https://www.nber.org/papers/w5896

Krugman, P. (1991). Increasing returns and economic geography. *Journal of Political Economy*, 99(3): 483–499.

Lai, Y.W., & Narayanan, S. (1999). Technology utilization level and choice: The electronics and electrical sector in Penang, Malaysia. JomoK, S., Felker, G. & Rasiah, R. (eds.), *Industrial Technology Development in Malaysia: Firm and Industry Studies*, London: Routledge, pp. 107–124.

Lall, S., & Siddharthan, N.S. (1982). The monopolistic advantages of multinationals: Lessons from foreign investment in the U.S. *Economic Journal*, 92: 668–683.

Lee-in, C.C. (2007). *The Policy, Institution and Market Factors in the Development of Taiwan's Textile/Garment Industry*. Chung-Hua Institution for Economic Research. Working Paper Series No.2007-2.

Levy, B., & Spiller, P. (eds.) (1996). *Regulations, Institutions, and Commitment: Comparative Studies of Telecommunications*. Cambridge: Cambridge University Press.

Lobova S., Bykovskaya, N.V., Vlasova, I.M., & Vlasova, I.M. (2019). Successful experience of formation of industry 4.0 in various countries. Popkova, E.G., Ragulina, Y.V. & Bokoviz, A.V. (eds.), *Industry 4.0: Industrial Revolution of the 21st Century*, New York: Springer, pp. 121–129.

Malaysia (2020). *Labour Market Review*. Malaysia, Putrajaya: Department of Statistics, Malaysia.

Matteucci, N., O' Mahony, M., Robinson, C., & Zwick, T. (2005). Productivity, workplace performance and ICT: Industry and firm-level evidence for Europe and the US. *Scottish Journal of Political Economy*, 52(3): 359–386.

MIDA (2020). *Key Highlights of the Manufacturing Sector in 2020, Malaysian Industrial Development Authority.* Accessed on June 14, 2022 from https://www.mida.gov.my/industries/manufacturing/)

MIDA (2022). *Building Technology & Lifestyle Division, Malaysian Industrial Development Authority.* Accessed on June 14, 2022 from https://www.mida.gov.my/staffdirectory/building-technology-lifestyle-division/

MITI (2020). *Digital Economy, Ministry of International Trade and Industry.* Accessed on June 14, 2022 from https://www.miti.gov.my/NIA/digital-economy.html

Nelson, R.R., & Winter, S.G. (1982). *An Evolutionary Theory of Economic Change.* Cambridge: Harvard University Press.

Newman, C., & Page, J.M. (2017). *Industrial clusters: The case for special economic zones in Africa* (No. 2017/15). WIDER Working Paper.

North, D.C. (1990). *Institutions, Institutional Change and Economic Performance.* Cambridge: Cambridge University Press.

Pérez, C. & Soete, L. (1988). Catching up in technology: Entry barriers and windows of opportunity. Dosi, G., Freeman, C., Nelson, R.R., Silverberg, G. & Soete, L. (eds.), *Technical Change and Economic Theory*, London: Pinter.

Pietrobelli, C., & Rabellotti, R. (2006). *Upgrading to Compete: Global Value Chains, Clusters, and SMEs in Latin America.* Cambridge: Inter-American Development Bank & Harvard University Press.

Porter, M.H. (1990) The Competitive Advantage of Nations, New York: Free Press.

Porter, M.E., & Stern, S. (2001). Innovation: location matters. *MIT Sloan Management Review*, 42(4): 28.

Rasiah, R. (1995). *Foreign Capital and Industrialization in Malaysia.* Basingstoke: Macmillan.

Rasiah, R. (2004). *UNU-INTECH Survey Questionnaire.* Maastricht: unpublished.

Rasiah, R. (2007). The systemic quad: Technological capabilities and economic performance of computer and component firms in Penang and Johor, Malaysia. *International Journal of Technological Learning, Innovation and Development*, 1(2): 179–203.

Rasiah, R. (2012). Beyond the multi-fibre agreement: How are workers in East Asia faring? *Institutions and Economies*, 4(3): 1–20.

Rasiah, R. (2019). Building networks to harness innovation synergies: Towards an open systems approach to sustainable development. *Journal of Open Innovation: Technology, Market, and Complexity*, 5(3): 70.

Rasiah, R., & Gachino, G. (2005). Are foreign firms more productive and export-and technology-intensive than local firms in Kenyan manufacturing? *Oxford Development Studies*, 33(2): 211–227.

Rasiah, R., & Lin, Y. (2005). Learning and innovation: The role of market, government and trust in the information hardware industry in Taiwan. *International Journal of Technology and Globalization*, 1(3/4): 400–432.

Rizzo, A., & Glasson, J. (2012). Iskandar Malaysia. *Cities*, 29(6): 417–427.

Robinson, P.K. & Hsieh, L. (2016). Reshoring: A strategic renewal of luxury clothing supply chains. *Operations Management Research*, 9: 89–101.

Saxenian, A.L. (1990). Regional networks and the resurgence of Silicon Valley. *California Management Review*, 33(1): 89–112.

Schumacher, A., Erol, S., & Sihna, W. (2016). A maturity model for assessing industry 4.0 readiness and maturity of manufacturing enterprises. *Procedia CIRP*, 52: 161–166.

Schumpeter, J. (1942). *Capitalism, Socialism, and Democracy*. Cambridge: Harvard University Press.

Schwab, K. (2017). *The Fourth Industrial Revolution*. Harmondsworth: Penguin.

Shaver, M.J., & Flyer, F. (2000). Agglomeration economies, firm heterogeneity, and foreign direct investment in the United States'. *Strategic Management Journal*, 21(12): 1175–1193.

Star (2021). TPM to develop new tech clusters, accessed on December 26, 2022 from https://www.thestar.com.my/business/business-news/2021/11/01/tpm-to-develop-new-tech-clusters

Taalbi, J. (2018). Origins and pathways of innovation in the third industrial revolution. *Industrial and Corporate Change*, 28(5): 1125–1148.

Vaidya, S., Prashant A., & Santosh, B. (2018). Industry 4.0—A glimpse. *Procedia Manufacturing*, 20: 233–38.

Wade, R.H. (1990). *Governing the Market: Economic Theory and the Role of Government in East Asian Industrialization*. Princeton: Princeton University Press.

World Bank (2017). *Knowledge in Action 2017: Trade & Competitiveness Publications Update*, Washington DC: World Bank.

6 Did Digitalization Help Manufacturers Cope with the Covid-19 Pandemic?

Shankaran Nambiar and Yip Tien Ming

Introduction

Malaysia has been actively promoting the digitalization of the economy. This has been done for several reasons, aside from the fact that Malaysia will lose out if it is not in tandem with other countries in the region that are also undertaking digitalization. First, the economy has been going through a long period of deindustrialization, and digitalization can lead to the future upward movement of Malaysian industries, particularly the export-oriented manufacturing sectors. Second, digitalization is expected to improve productivity in the economy, specifically in manufacturing. One of the problems that the country faces is the fact that wages have not been growing commensurate with the growth of the economy. Thus, digitalization can lead to a push in wage growth rates. Third, there is a heavy dependence on foreign labour. It is expected that digitalization will bring about job disruption, removing the need for unskilled and semi-skilled migrant labour. Digitalization will also bring about job creation, leading to the creation of jobs that require more skilled labour and those that are suited to meet the challenges of a knowledge economy. Finally, digitalization is also necessary for the development of the economy and to bring up to the border of the production possibility curve through the introduction of the Industrial Revolution 4.0 (IR4.0). This will encompass robotics, the internet of things (IoT), blockchain technologies and greater automation. Malaysia will need to jump up the value chain through IR4.0 if it is to escape the middle-income trap and emerge as one of the leaders in ASEAN.

There is a broader context within which digitalization is being introduced to the economy. The Twelfth Malaysia Plan (12MP), a document that has been formulated to drive the Malaysian economy for the next five years, rests on two drivers of growth – sustainability and digitalization. An important aspect of the 12MP is to bring about a post-Covid recovery of the economy, and it is in this respect that digitalization has an important role to play. The 12MP plans to transform Malaysia into a technology-based economy with a prominent role in the growth of the digital economy. It is predicted that under the plan period, the digital economy will contribute 25.5 per cent of the GDP with digital content exports achieving an average annual growth rate of 8 per cent. The report

DOI: 10.4324/9781003367093-6

also revealed that Malaysia will capitalize on advanced technology potential by gearing up for the Fourth Industrial Revolution.

The Covid-19 pandemic caught the Malaysian industry by surprise. Manufacturers were not prepared for the pandemic and had to improvise quickly to minimize the damage caused. Some of the problems that they had to face included the shortage of foreign labour, the imposition of standard operation procedures (SOPs) which reduced the number of workers available at factories at any one time, and the disruption of logistics. There were other problems such as congestion at ports and the bottlenecks in obtaining imported inputs.

Manufacturers in Malaysia responded quickly to the effects of the pandemic, which included the more extensive use of electronic commerce, e-platforms and flexible delivery methods. However, not all companies were able to respond quickly and effectively because many companies had not yet embraced even 4G, nor were they familiar with automation, robotics and artificial intelligence (AI). Nevertheless, the Covid-19 pandemic period with the regular movement control orders (MCOs) that restricted movement made it clear that digitalization was inevitable. This period demonstrated the necessity of digitalization and underscores the point that Malaysia has to pursue digitalization with a sense of urgency. It is in this context that this chapter has an important contribution to make. It points out that digitalization is essential and discusses the Malaysian case where manufacturers have mitigated the effects of the pandemic through digitalization.

Despite the important role of digitalization in helping manufacturers to cope with the Covid-19 pandemic, empirical investigation on this subject matter is limited in the existing literature. To the best of our knowledge, Guo et al. (2020) were pioneers in investigating the effect of digitalization on firms' performance in China during the Covid-19 pandemic period. The study found that highly digitalized firms are associated with better crisis response strategies, thereby mitigating the adverse impact of Covid-19 on the firm performance. The mitigating role of digitalization is further reaffirmed by Belhadi et al. (2021) for firms in a group of developed countries.

In light of the limited empirical evidence, this study aims to investigate whether digitalization mitigates the negative impact of the Covid-19 pandemic on Malaysia's manufacturing sector performance. This study addresses three main drawbacks in the existing literature. First, empirical study on the mitigating role of digitalization primarily focused on China and developed countries. The findings obtained may not be generalized in Malaysia due to the different levels of digital adoption. As indicated earlier, many manufacturers had not yet embraced even 4G, nor were they familiar with automation, robotics and artificial intelligence (AI). Consequently, this would render the role of digitalization in helping manufacturers to cope with the Covid-19 pandemic. Hence, this study attempts to provide a country-specific analysis of Malaysia to validate the mitigating role of digitalization. Second, existing studies by Guo et al. (2020) and Belhadi et al. (2021) employed survey data to quantify the mitigating effect of digitalization.

However, survey data may be subject to response bias which would question the accuracy of the estimation results. Meanwhile, this study complements the existing studies by using secondary data published by the Malaysian authority to study the mitigating role of digitalization. For instance, the manufacturing industrial production index, which was used as a proxy for manufacturing performance in this study, was calculated based on the actual production value of all the industries in a particular period. The data is free from response bias as it is collected directly from the statement of production of the manufacturers. Third, Guo et al. (2020) employed a cross-tabulation method to quantify the mitigation effect of digitalization. Meanwhile, this study put forth an alternative model specification by providing the marginal effect of the Covid-19 pandemic on manufacturing performance at various levels of digitalization.

Based on a balanced panel data of 130 manufacturing industries in the Malaysian manufacturing sector over the period January 2020 to December 2020, the empirical findings validate the mitigating role of digitalization in the context of the manufacturing sector in Malaysia. The results show that an additional increase in the Covid-19 cases per million of the population is associated with a 0.095 percent reduction in the manufacturing industrial production performance when the country's digital adoption is at the minimum level (2182 thousands of Telekom Malaysia's fixed broadband subscribers). However, the adverse impact of the Covid-19 pandemic diminishes with the rising level of digital adoption. When digital adoption is at the maximum level (2366 thousand of Telekom Malaysia's fixed broadband subscribers), the negative impact reduces to 0.001 percent. This is equivalent to a 98 percent reduction in the negative effect of the Covid-19 pandemic. The effect is substantial, implying that digitalization plays an essential role in helping the manufacturers to cope with the Covid-19 pandemic. More importantly, the Covid-19 pandemic has an insignificant influence on the manufacturing industrial production performance when the level of digital adoption is at the maximum level. Taken together, the results highlight the important role of digitalization in offsetting the adverse impact of the Covid-19 pandemic on the manufacturing sector's performance

This chapter is organized as follows. The next section provides a theoretical review. This is followed by a discussion of the data and empirical model used. Section "Data and Empirical Mode" deals with the empirical results. Section "Findings and Discussion" is concerned with the robustness checks. Finally, some concluding remarks are offered.

Theoretical Considerations

This study builds on the firm-level study by Guo et al. (2020), which argues that digitalization plays a mitigating role and can offset the adverse impact of the Covid-19 pandemic on firms' performance. The study points out that digitalization provides effective crisis response strategies to cope with the pandemic. Specifically, highly digitalized firms are associated with greater dynamic capabilities in terms of sensing the crisis, seizing new opportunities in the crisis and reconfiguring resources to cope with the crisis. This would result in effective crisis responses strategies and subsequently offset the adverse impact of the pandemic on firms' performance.

Dynamic capabilities refer to the firms' capability to build, integrate and reconfigure resources when coping with a rapidly changing environment (Treece et al., 1997; Treece, 2007; 2012). This perspective is the key to formulating effective crisis response strategies for firms, wherein firms with dynamic capabilities are more likely to adapt to the changing environment and sustain company operations during the crisis period (Eisenhardt and Jeffrey, 2000 and Lin and Wu, 2014).

In the context of a crisis, dynamic capabilities involve three dimensions, namely the capability of sensing the crisis, the capability of seizing new opportunities in the crisis and the capability of reconfiguring resources to cope with the crisis (Treece, 2007; Ballesteros et al., 2017; Guo et al., 2020). The first dimension indicates the ability to sense or understand the crisis in a timely manner. Firms with poor dynamic capabilities would be unable to predict the arrival of an unprecedented event, such as the outbreak of the Covid-19 pandemic. However, with dynamic capabilities, firms are able to gauge the possible implication of the pandemic on the company's operation, such as production interruption, supply chain distortion and order cancellation. Hence, firms with dynamic capabilities would be better able to perceive a crisis and subsequently develop comprehensive strategies to overcome the challenges imposed by the pandemic.

The second dimension posits that firms with dynamic capabilities are more likely to seize new business opportunities, and subsequently provide an alternative business model to help the companies sustain themselves during a crisis period. In applying to the Covid-19 outbreak, the pandemic has resulted in the closure of physical stores and the implementation of social distancing measures, which led consumers to switch from offline to online purchases. This created new business opportunities for firms, in particular the e-commerce channels. Therefore, firms with dynamic capabilities would leverage the online platform to market their products and services, thereby preserving firms' sales performance during the pandemic period.

The third dimension proposes that firms with dynamic capabilities are able to integrate internal and external resources to cope with the crisis. Applying this to the Covid-19 outbreak, the pandemic has resulted in plant closures, especially in the manufacturing sector in order to curb the virus transmission. The shutdown in operation is costly and subsequently has hampered the firms' financial position. Hence, firms with dynamic capabilities are more likely to employ external resources to ensure the continuation of business operations. For instance, firms may equip employees with digital technologies such as remote robots in conducting the production activity. This allows the continuation of production activity despite the shutdown of plants and subsequently offsets the revenue losses caused by the pandemic.

Meanwhile, digitalization refers to the process of organizational transformation through the adoption of digital technologies (Sebastian et al., 2017; Vial,

2019; Guo et al., 2020). Digitalization has been widely perceived to be the key to sustainable development for business enterprises (OECD, 2014). Digitalization, particularly the application of big data analysis, enables firms to predict the changing environment to some extent, thereby allowing them to have a better gauge of the country's future economic prospects. This assists firms in designing short- and long-term strategies in sustaining companies amid future economic uncertainty. Moreover, digitalization allows firms to venture into new business models, such as the digital business to expand their consumer market. Furthermore, digitalization enables the use of robots in the production process, thereby improving production efficiency and allowing firms to achieve greater economies of scale.

In relation to this study, digitalization would enhance the dynamic capabilities of manufacturers, thereby providing them with a better response to the Covid-19 pandemic and subsequently preserve firms' performance (Guo et al., 2020). In a broader context, digitalization is expected to offset the adverse impact of the Covid-19 pandemic on manufacturing sector performance.

For the first dimension in the dynamic capabilities, digitalization would enhance manufacturers' capability in sensing the crisis. Digitalization allows them to identify the spread of Covid-19, thereby providing an early indication on the possible implication of the pandemic on firms' performance. Subsequently, early preparation can be conducted to cope with the pandemic. For instance, the use of big data analysis would help manufacturers to predict environmental changes to some extent (George et al., 2014) and allows them to have a better perception of the pandemic. This in turn helps them to formulate strategies to cope with the pandemic.

As far as the second dimension is concerned, manufacturers can better seize business opportunities with the help of digital technology. Applying this dimension to the Covid-19 outbreak, digitalization enables firms to venture into a digital business model by providing an online platform to market their products and services. This expands the consumer market and provides a constant revenue stream to the company, offsetting the revenue loss due to the closure of physical stores during the pandemic period.

The third dimension, as we have discussed, refers to the integration of resources to respond to crises. The crisis was a challenge for firms because manufacturers had to face many challenges, such as staff shortages, production disruptions and delays in launching new products. There are various ways in which resources could have been used to mitigate the effects of the pandemic. For instance, digital technology such as remote robots would have helped manufacturers in continuing their output production. Factories equipped with remote control robots could have been used to perform their daily routine in the factories. Hence, this ensures the continuation of manufacturing activity despite the closure of plants. Thus, manufacturers who adopt a high degree of digitalization are more likely to be able to quickly reshape their operation and subsequently minimize the adverse impact of the pandemic.

It can be argued that digitalization can play an important role in moderating the relationship between the Covid-19 pandemic and manufacturing sector performance. In particular, the Covid-19 pandemic is conjectured to depress the manufacturing sector's performance through the reduction in output produced and purchase orders. Arguably, digitalization would mitigate these adverse effects by providing effective crisis response strategies to the manufacturers, enabling them to cope with the pandemic. Specifically, highly digitalized firms are associated with greater dynamic capabilities in terms of sensing the crisis, seizing new opportunities in the crisis and reconfiguring resources to cope with the crisis. This would result in effective crisis responses strategies and subsequently offset the adverse impact of the pandemic on firms' performance, and vice versa.

Data and Empirical Model

The data and the empirical model specified to answer the question we began with are discussed in this section.

Data

This study employs a balanced panel data of 130 manufacturing industries in the Malaysian manufacturing sector over the period January 2020 to December 2020. The industry classification is based on the Malaysia Standard Industrial Classification (MSIC) 2000, whereby the manufacturing sector is categorized into 130 industries based on the five-digit code. Industry-level data is used instead of firm-level data because the former is available on monthly frequency while the latter is available on an annual basis. Hence, the use of industry-level data permits sufficient sample periods to quantify the impact of the Covid-19 pandemic on the manufacturing sector performance.

The main dependent variable in this study is the yearly growth rate of the manufacturing industrial production index ($IPIYoY$), a proxy for the manufacturing sector's performance. Next, the Covid-19 pandemic is proxied by the total Covid-19 cases per million population ($COVID$) and digital adoption ($DIGITAL$) is proxied by the number of fixed broadband subscribers of Telekom Malaysia. There are two reasons for choosing this proxy. First, Telekom Malaysia is the leading internet service provider in Malaysia, and therefore the number of fixed broadband subscribers is likely to capture a broad range of internet users as compared to other internet service provider companies such as Maxis, Digi, Celcom and U Mobile. Hence, it can be expected that the number of Telekom Malaysia's fixed broadband subscribers is likely to capture the pace of digitalization in Malaysia.

Second, broadband penetration has been recognized by the World Bank as an important source of growth for a country (Minges, 2016). As such, broadband enables services such as cloud computing and mobile apps, which facilitate innovations across various sectors of the economy. With respect to the present study, broadband internet could be expected to facilitate innovation and productivity in the manufacturing sector by digitalizing the production process. Digitalization enables the use of robots and subsequently facilitates the production process, allowing projects to move faster and manufacturers to receive more orders. As a result, digitalization enables manufacturers to practice economies of scale, thereby improving the output produced and company performance.

However, the number of fixed broadband subscribers of Telekom Malaysia is available at a quarterly frequency. Hence, this study employs the cubic spline interpolation method to convert the quarterly subscribers into monthly frequency. This interpolation method is discussed in detail in Granville (2005) and is widely adopted in empirical studies to overcome data unavailability (Bathia and Bredin, 2013; Isabel, 2017; Shaharuddin et al., 2018).

Additional control variables such as gross export (*EXPORT*), the exchange rate (*MYRUSD*) and industrial production index (*IPI*) are included to avoid omitted variable bias in the model. The inclusion of the variables allows the model to capture the impact of external demand, exchange rate variation and country business cycle effects on the manufacturing sector performance, respectively. Next, to ensure the result is robust against alternative measures of Covid-19 pandemic and digitalization proxy, the new Covid-19 cases per million population (*COVIDNEW*), internet banking penetration rate (*IB*) and mobile banking penetration rate (*MOBILE*) are included in the robustness check section to avoid the measurement error. All the variables used in the baseline estimation and robustness check are shown in Table 6.1.

Empirical Model

The empirical model use to examine the mitigating effect of digitalization is shown as Eq. (1).

Table 6.1 List of Variables

Variables	Description	Unit of Measurement	Source
Variables use in baseline model			
Ipiyoy	Manufacturing industrial production index	Year-on-year percent	DOSM
Covid	Total Covid-19 cases	Per million population	Our World in Data
Digital	Telekom Malaysia's fixed broadband subscriber	Thousands	Telekom Malaysia
Export	Gross export	Million MYR	BNM
Myrusd	Dollar to Ringgit exchange rate	Exchange rate	BNM
Ipi	Industrial production index	Index	BNM
Variables use in robustness checks			
Covidnew	New Covid-19 cases	Per million population	Our World in Data
IB	Internet banking penetration rate	Percent of total population	BNM
MOBILE	Mobile banking penetration rate	Percent of total population	BNM

Note: DOSM denotes Department of Statistics Malaysia, BNM denotes Bank Negara Malaysia.

$$IPIY_oY_{it} = \beta_0 + \beta_1 IPIY_oY_{it-1} + \beta_2 COVID_t + \beta_3 DIGITAL_t \tag{1}$$
$$+\beta_4 COVID_t \text{x} DIGITAL_t + \beta_5 LnEXPORT_t$$
$$+\beta_6 LnMYRUSD_t + \beta - LnIPI_t + v_i + \varepsilon_{it}$$

where i is the industry index, t is the month index, *IPIYoY* represents the yearly growth rate of the manufacturing industrial production index, *COVID* represents the total Covid-19 cases per million population and *DIGITAL* denotes digital adoption, proxy by the number of fixed broadband subscribers of Telekom Malaysia. Both Covid-19 and digital adoption measures are not available by industry level, and therefore aggregate value is included into the model. Despite that, the use of aggregate value is able to capture the impact of Covid-19 and digital adoption on manufacturing performance. For instance, higher Covid-19 cases induced the Malaysian government to implement the national lockdown policy to curb the virus' transmission. Consequently, this halted manufacturing activity, whereby factories were not allowed to operate during the lockdown period. This in turn negatively affected the performance of the manufacturing industry.

For the digital adoption measure, a higher number of fixed broadband subscribers indicates a greater degree of internet penetration rate in the country, thereby facilitating manufacturing performance. For the consumer, the use of the internet allows them to purchase through the e-commerce platform. This ensures consistent demand for manufacturers' goods amid the closure of physical stores. For the manufacturers, the use of the internet allows firms to venture into digital business models by providing an online platform to market their products and services. This would expand the consumer market and provides a constant revenue stream to the company, thereby offsetting the revenue loss due to the closure of physical stores during the pandemic period. More importantly, the use of the internet, in particular high-speed internet, such as 5G network, allows manufacturers to deploy remote robots in the factory. The use of remote robots enables employees to perform their daily routine in the factory, thereby ensuring the continuation of manufacturing activity despite the closure of the plant. This facilitates production, whereby the output continues to be produced and the performance of the manufacturing sector is maintained.

Among the control variables, *EXPORT* represents gross exports, *MYRUSD* denotes the Dollar to Ringgit Malaysia exchange rate and *IPI* represents the industrial production index. The inclusion of the variables allows the model to capture the impact of external demand, exchange rate variation and country business cycle effects on the manufacturing sector performance, respectively. Here v_i represents the industry-specific effect, ε_{it} represents the error term and Ln denotes natural logarithm. Next, the inclusion of the lagged dependent variable is to control for the dynamic effect in manufacturing activity. Current period manufacturing performance is highly affected by the previous period's performance. This is pertinent during the Covid-19 pandemic, which was a period of sluggish manufacturing performance; it was subsequently associated with a rebound in the manufacturing activity in the following periods. Evidently, the yearly growth rate of the overall manufacturing industrial production index was negative 33.2 percent in April 2020, which was when the MCO started. However, the index performance rebounded in May 2020 with a positive yearly growth rate of 22.7 percent.

As noted by Brambor et al. (2006), it is inappropriate to interpret the estimated coefficients of Covid-19 (*COVID*) and digital adoption (*DIGITAL*) if the model contains an interaction term (*COVIDxDIGITAL*), as the former captures the effect of Covid-19 pandemic on manufacturing sector performance only if the digital adoption is equal to zero. The latter provides the effect of digital adoption on manufacturing sector performance only if the Covid-19 cases are equal to zero. Hence, the coefficients of *COVID* and *DIGITAL* are meaningless with the inclusion of interaction terms in the model. Furthermore, it is also inappropriate to interpret the estimated coefficients of the interaction term β_4 as this would result in misstating bias as indicated by Kingsley et al. (2017). To interpret the result, this study follows Brambor et al. (2006) and Kingsley et al. (2017) suggestion by computing the marginal effect of the Covid-19 pandemic on manufacturing sector performance at different levels of digital adoption. That is:

$$\frac{\partial IPIY_{o}Y_{it}}{\partial COVID_{t}} = \beta_2 + \beta_4 DIGITAL_t s \qquad (2)$$

The Fixed Effect (FE) with robust standard error estimator is used to estimate Eq. (1). However, the presence of lagged dependent variable would result in Nickell's (1981) bias, whereby the bias arises from the non-zero correlation between the lagged dependent variable and the industry-specific effect. This in turn affects the estimated coefficient of the lagged dependent variable. Despite that, the bias in the slope coefficients is marginal as demonstrated by Judson and Owen (1999). Given that the focus of this study is on the slope coefficients instead of the lagged dependent variable, therefore the FE estimator is appropriate in the context of this study. For concreteness, this study also employs the bias-corrected LSDV (LSDVC) to correct for the bias inherent in the estimated coefficient for the lagged dependent variable. The estimation is conducted based on the *xtlsdvc* STATA routine developed by Bruno (2005). As demonstrated by Kiviet (1995), Judson and Owen (1999) and Bun and Kiviet (2003), the LSDVC estimator performs better than the Anderson-Hsiao Instrumental variable and GMM estimator (first-difference and system) in terms of bias and root mean squares errors for both balanced and unbalanced panel, respectively. Hence, the LSDVC estimator is the preferred estimator to mitigate Nickell's bias in the context of this study.

Findings and Discussion

We examine the descriptive statistics and analytic findings separately. The discussion ensuring from that is undertaken in the findings sub-section.

Descriptive Statistics

Table 6.2 shows the descriptive statistics for all the variables used in this study. As observed, the yearly growth rate of the manufacturing industrial production index was negative 4.02 percent, indicating that the outbreak of the Covid-19 pandemic has halted the manufacturing sector's performance as compared to the preceding year. Next, on average, the total Covid-19 cases stood at 682 cases per million population, far surpassing neighbouring countries such as Thailand (41 total Covid-19 cases per million population), Brunei (328 total Covid-19 cases per million population) and Vietnam (6 total Covid-19 cases per million population). Meanwhile, the average number of Telekom Malaysia's fixed broadband subscribers stood at 2253 thousand in 2020.

Baseline Results

Table 6.3 shows the estimation results for Eq. (1). As observed, the lagged dependent variable is significant at 1 percent level, indicating the presence of dynamic effects in the Malaysian manufacturing activity. Hence, the inclusion of lagged dependent variable is justifiable and allows the model to avoid omitted variable bias. Turning to the focus of this study, the marginal effect estimates in both specifications show the mitigating role of digitalization. Focusing on specification 1, the Covid-19 pandemic has a negative and significant impact on manufacturing industrial production performance if the digital adoption is at the minimum level (2182 thousand of Telekom Malaysia's fixed broadband subscriber).

The negative effect is estimated at 0.095 percent. However, the adverse impact of the Covid-19 pandemic diminishes with the rising level of digital adoption. When digital adoption is at the maximum level (2366 thousand of Telekom Malaysia's fixed broadband subscribers), the negative impact reduces to 0.001 percent. This is equivalent to a 98 percent reduction in the negative effect of the Covid-19 pandemic. The effect is substantial, implying that digitalization plays an essential role in helping the manufacturers to cope with the Covid-19 pandemic. More importantly, the Covid-19 pandemic has insignificant influence on

Table 6.2 Descriptive Statistics

Variable	Mean	Std. dev	Obs
Ipiyoy	−4.023	23.468	1560
Covid	682.439	1006.935	1560
Digital	2253.260	56.916	1560
export	81726.330	9927.055	1560
myrusd	4.198	0.103	1560
ipi	109.774	12.330	1560
Covidnew	12.258	22.613	1560
Ib	105.890	4.749	1560
Mobile	56.877	2.9230	1560

Note: Sample period: January 2020–December 2020.

the manufacturing industrial production performance when the level of digital adoption is at the maximum level. Taken together, the result highlights the important role of digitalization in offsetting the adverse impact of the Covid-19 pandemic on the manufacturing sector performance.

Similar findings are documented in specification 2 where LSDVC is the estimation method. The finding on the mitigating effect of digitalization is in line with the theoretical argument put forth by Guo et al. (2020), whereby digitalization is essential in improving firms' dynamic capabilities and subsequently provides effective response strategies to cope with the Covid-19 pandemic. Empirically, the finding concurs with the study by Belhadi et al. (2021), in which digital adoption is essential in helping firms to overcome the challenges imposed by the Covid-19 pandemic and sustain their business operations.

Table 6.3 Baseline Estimation Result

Specifications	1	2
Estimation Methods	FE	LSDVC
Ipiyoyit–1	0.35204***	0.50399***
	(0.02981)	(0.26007)
Covidit	−0.10723**	−0.15978**
	(0.04274)	(0.07147)
Digitalit	0.08543***	0.14898***
	(0.02887)	(0.03041)
Covidit*digitalit	0.00004**	0.00007**
	(0.00002)	(0.00003)
Lnexportit	−55.96104***	−72.21403***
	(6.91714)	(8.34966)
lnmyrusdit	115.95940**	250.79440***
	(47.71059)	(52.65170)
LnIPIit	122.69770***	138.29290***
	(9.71720)	(7.54477)
Constant	−302.87280***	–
	(105.12680)	
Marginal effect		
Minimum	−0.09459**	−0.01386**
	(0.00385)	(0.00615)
Mean	−0.00628**	−0.00911**
	(0.00261)	(0.00406)
Maximum	−0.00119	−0.00153
	(0.00078)	(0.00107)
No. industries	130	130
Observation	1430	1430

Note:
1 *, ** and *** denote statistically significant at 10 percent, 5 percent and 1percent level, respectively.
2 Values in parentheses are robust standard error. Bootstrap standard error is reported for LSDVC.
3 Due to lagged dependent variable, the data for January 2020 is effectively drop out for all the industries. Therefore, the final observation is 1430.
4 Ln denotes natural logarithm.

Robustness Checks

Alternative Measures of Covid-19 Cases

A series of robustness checks are conducted to ensure the validity of the finding on the mitigating effect of digitalization. Table 6.4 shows the robustness check results with alternative measures of the Covid-19 pandemic, namely the new Covid-19 cases per million population (*COVIDNEW*). Notably, the mitigating effect of digitalization is consistent in all the specifications, in which the negative influence of Covid-19 pandemic on manufacturing industrial production performance deteriorates with higher level of digital adoption. This indicates the baseline finding is robust and not influenced by alternative measures of Covid-19 cases.

Table 6.4 Robustness Check with New Covid-19 Cases Per Million Population

Specifications	1	2
Estimation Methods	FE	LSDVC
Ipiyoyit-1	0.36474***	0.52333***
	(0.03025)	(0.02637)
Covidnewit	−7.42609***	−10.77697***
	(1.51469)	(2.21307)
Digitalit	0.07378***	0.13012***
	(0.02268)	(0.02160)
Covidnewit*digitalit	0.00313***	0.00455***
	(0.00064)	(0.00094)
Lnexportit	−55.53618***	−72.55968***
	(6.77094)	(8.01679)
lnmyrusdit	106.72500**	242.36900***
	(44.69958)	(49.81513)
Ln1Plit	122.39570***	137.66930***
	(9.82484)	(7.44767)
Constant	−267.30670***	–
	(97.85287)	
Marginal effect		
Minimum	−0.60539***	−0.85441***
	(0.13063)	(0.17494)
Mean	−0.38338***	−0.53144***
	(0.08644)	(0.11057)
Maximum	−0.02916	−0.01613
	(0.02428)	(0.03442)
No. industries	130	130
Observation	1430	1430

Note:
1 *, ** and *** denotes statistically significant at 10 percent, 5 percent and 1 percent level, respectively.
2 Values in parentheses are robust standard error. Bootstrap standard error is reported for LSDVC.
3 Due to lagged dependent variable, the data for January 2020 is effectively drop out for all the industries. Therefore, the final observation is 1430.
4 Ln denotes natural logarithm.

Alternative Measures of Digital Adoption

This robustness check considers alternative proxies for digital adoption, namely the internet and mobile banking penetration rate. A high degree of internet or mobile banking penetration rate implies greater usage of e-commerce for the purchase of goods and services. Subsequently, this would induce firms or businesses to venture into digital business to expand their consumer market. Arguably, a higher level of internet or mobile banking penetration rate is expected to be associated with a higher degree of digital adoption among the firms. Thus, both

Table 6.5 Robustness Check with Alternative Measures of Digital Adoption

Specifications	1	2	3	4
Estimation Methods	FE	LSDVC	FE	LSDVC
Ipiyoyit–1	0.36111***	0.52009***	0.37886***	0.53326***
	(0.02951)	(0.02633)	(0.02831)	(0.02571)
Covidit	−0.28065***	−0.36256***	−0.27890***	−0.31851***
	(0.07125)	(0.10526)	(0.05201)	(0.06735)
Ibit	0.93107***	1.46093***	–	–
	(0.23305)	(0.22956)		
Covidit*IBit	0.00249***	0.00324***	–	–
	(0.00063)	(0.00094)		
Mobileit	–	–	3.69848***	4.99413***
			(0.62265)	(0.60165)
Covidit*mobileit	–	–	0.00452***	0.00517***
			(0.00084)	(0.00109)
Lnexportit	−51.81035***	−65.80945***	−57.21500***	−71.05632***
	(6.33355)	(7.53777)	(6.66017)	(7.38841)
Lnmyrusdit	140.08040***	281.78450***	453.89350***	665.23720***
	(42.20184)	(53.34006)	(77.40898)	(82.26044)
LnIPIit	111.92530***	122.12210***	124.11720***	138.47700***
	(9.85158)	(7.38432)	(9.95486)	(7.17482)
Constant	−239.34590***	–	−796.69740***	–
	(71.80774)		(126.01550)	
Marginal effect				
Minimum	−0.03459***	−0.04381***	−0.04026***	−0.04523***
	(0.00899)	(0.01315)	(0.00772)	(0.00993)
Mean	−0.01622***	−0.02000***	−0.02197***	−0.02428***
	(0.00435)	(0.00629)	(0.00434)	(0.00556)
Maximum	0.00019	0.00124	0.00049	0.00143*
	(0.00058)	(0.00083)	(0.00059)	(0.00084)
No. industries	130	130	130	130
Observation	1430	1430	1430	1430

Note:
1 *, ** and *** denotes statistically significant at 10 percent, 5 percent and 1percent level, respectively.
2 Values in parentheses are robust standard error. Bootstrap standard error is reported for LSDVC.
3 Due to lagged dependent variable, the data for January 2020 is effectively drop out for all the industries. Therefore, the final observation is 1430.
4 Ln denotes natural logarithm.

measures can be regarded as the alternative proxy for digitalization. Table 6.5 shows the estimation result using the internet and mobile banking penetration rate. Consistent with the baseline finding, the marginal effect estimate shows that the negative impact of the Covid-19 pandemic on manufacturing industrial production performance reduces with the rising level of internet and mobile banking penetration rate.

Conclusion

This study argues that the Malaysian government should recognize the importance of digitalization for the growth of the economy. Therefore, the government has prioritized digitalization as part of the national agenda, emphasizing it in the 12MP. The Covid-19 outbreak lay open the gaps in the country's preparedness to handle a public health crisis such as that caused by the virus. The manufacturing sector was also badly hit by the crisis since there was a domestic, as well as an international, dimension to the crisis.

The theoretical framework indicates that firms need to have dynamic capabilities in order to handle a crisis – and there is no doubt that the Covid-19 pandemic was nothing less than a public health crisis. Covid-19 presented firms with the challenge of building, integrating and reconfiguring resources in a rapidly changing environment such as the health crisis presented. We argue that firms, in the context of dynamic capabilities, must have the capability of sensing the onset of a crisis as a first step. Next, they must be able to figure out new opportunities in dealing with the crisis. Finally, firms must have the capacity to reconfigure resources to handle the crisis.

The empirical model that we have used indicates that the growth rate of the manufacturing sector depends, among other variables, on digital adoption, gross exports, the exchange rate and the incidence of Covid rates. Our results indicate that digital adoption mitigates the adverse impact of Covid-19 pandemic on the manufacturing sector performance. This is an encouraging result because it indicates that digitalization has the possibility of helping the manufacturing sector withstand shocks such as Covid-19 where the public health system will be impacted and there will be movement restrictions.

There are, however, two problems. The first is the uptake of digitalization. The bigger firms and MNCs will take readily to digitalization; there will be no need to convince these firms of the need to digitalize. The constraint arises in so far as the smaller firms are concerned because they may not have the capital to invest in new technology. The other bottleneck lies in the availability of cheap foreign labour. As long as migrant labour is easily available, there will be little incentive to find substitutes to it. This will discourage firms from introducing digitalization in their factories.

The state has a role in the introduction of digitalization. Some of the policy measures that can be introduced include incentivizing firms to adopt digitalization. This will include fiscal measures, but it can also include measures to deincentivize the reluctance to adopt digitalization. The most obvious measure in this respect is to avoid the use of unskilled or semi-skilled migrant labour, which will dissuade firms from digitalizing. The availability of cheap migrant labour can counteract attempts to introduce digitalizing because with cheap labour freely available there will be no incentive to switch to more capital intensive methods.

Although in the long-run digitalization is useful for growth and development, firms may not see the need to do so immediately. Firms will typically be concerned with short-run costs and immediate returns rather than long-term efficiency. It is for this reason that state support is necessary to encourage firms to adopt digitalization.

References

Ballesteros, L., Useem, M. and Wry, T. (2017). Masters of disasters? An empirical analysis of how societies benefit from corporate disaster aid. *Academy of Management Journal, 60*(5), 1682–1708.

Bathia, D. and Bredin, D. (2013). An examination of investor sentiment effect on G7 stock market returns. *The European Journal of Finance, 19*(9), 909–937.

Belhadi, A., Kamble, S., Jabbour, C.J.C., Gunasekaran, A., Ndubisi, N.O. and Venkatesh, M. (2021). Manufacturing and service supply chain resilience to the COVID-19 outbreak: Lessons learned from the automobile and airline industries. *Technological Forecasting & Social Change, 163*, DOI: https://doi.org/10.1016/j.techfore.2020.120447.

Brambor, T., Clark, W.M. and Golder, M. (2006). Understanding interaction models: Improving empirical analyses. *Political Analysis, 14*, 63–82.

Bruno, G.S.F. (2005). Estimation and inference in dynamic unbalanced panel-data models with small number of individuals. *The Stata Journal Promoting Communications on Statistics and Stata, 5*(4), 473–500.

Bun, M.J.G. and Kiviet, J.F. (2003). On the diminishing returns of higher order terms in asymptotic expansions of bias. *Economics Letters, 79*(2), 145–152.

Eisenhardt, M. and Jeffrey, A.M. (2000). Dynamic capabilities: What are they? *Strategic Management Journal, 21*(10–11), 1105–1121.

George, G., Haas, M.R., and Pentland, A. (2014). Big data and management. *Academy Management Journal, 57*(2), 321–326.

Guo, H., Yang, Z., Huang, R. and Guo, A. (2020). The digitalization and public crisis response of small and medium enterprises: Implications from the COVID-19 survey. *Frontiers of Business Research in China.* DOI: https://doi.org/10.1186/s11782-020-00087-1

Granville, S. (2005). *Application: Cubic Spline Interpolation, Computational Methods of Linear Algebra*. Hoboken, NJ: Wiley-Interscience.

Isabel, V. (2017). *Did the Crisis Permanently Scar the Portuguese Labour Market? Evidence from a Markov-switching Beveridge Curve Analysis*. Working Papers No. 2043. European Central Bank.

Judson, R.A. and Owen, A.L. (1999). Estimating dynamic panel data models: A guide for macroeconomists. *Economics Letters 65*, 9–15.

Kingsley, A.F., Noordewier, T.G. and Bergh, R.G.V. (2017). Overstating and understating interaction results in international business research. *Journal of World Business, 52*(2), 286–295.

Kiviet, J.F. (1995). On bias, inconsistency, and efficiency of various estimators in dynamic panel data models. *Journal of Econometrics, 68*(1), 53–78.

Lin, Y. and Wu, L.Y. (2014). Exploring the role of dynamic capabilities in firm performance under the resource-based view framework. *Journal of Business Research, 67*(3), 407–413.

Minges, M. (2016). *Exploring the Relationship Between Broadband and Economic Growth.* World Development Report No. 102955. World Bank.

Nickell, S. (1981). Biases in Dynamic Models with Fixed Effects. *Econometrica, 49*(6), 1417–1426.

OECD (2014). The digital economy, new business models and key features. In *Addressing the Tax Challenges of the Digital Economy* (pp. 69–97), Paris: Organization for Economic Cooperation and Development (OECD).

Sebastian, I., Ross, J., Beath, C., Mocker, M., Moloney, K. and Fonstad, N. (2017). How big old companies navigate digital transformation. *MIS Quarterly, 16*(3), 197–213.

Shaharuddin, S.S., Lau, W.Y. and Ahamd, R. (2018). Is the Fama French Three-Factor model relevant? Evidence from Islamic unit trust funds. *Journal of Asian Finance, Economics and Business, 5*(4), 21–34.

Treece, D.J. (2007). Explicating dynamic capabilities: The nature and microfoundations of (sustainable) enterprise performance. *Strategic Management Journal, 28*(13), 1319–1350.

Treece, D.J. (2012). Dynamic capabilities: Routines versus entrepreneurial action. *Journal of Management Studies, 49*(8), 1395–1401.

Treece, D.J., Pisano, G. and Shuen, A. (1997). Dynamic capabilities and strategic management. *Strategic Management Journal, 18*(7), 509–533

Vial, G. (2019). Understanding digital transformation: A review and a research agenda. *The Journal of Strategic Information Systems, 28*(2), 118–144.

7 Creative Digital Pedagogies for Student Engagement

Preparing Students for Industry 4.0

Dorothy DeWitt and Norlidah Alias

Introduction

The rapid advances in technology supported by high-internet speeds and networks among people, systems and physical objects heralds the advent of the Fourth Industrial Revolution (IR4.0). The effects of the new technologies, such as automation, internet of things (IoT), analysis and big data, augmented and virtual reality (VR), as well as robotics and cloud utilization have already begun to impact industry (Schwab, 2016). In the manufacturing and economic sectors, these new technologies are being employed to transform Malaysia's economy (Ministry of International Trade and Industry, 2019). In the future of work after COVID-19, McKinsey reasons that remote work, digitalization and automation have shifted the labour demands across occupations (Lund, Madgavkar, Manyika, Smit, Ellingrud, Meaney & Robinson, 2021). Thus, digital and technical skills are important in the new market economies in the future of work. However, digital and technical skills are not the only skills needed. The lack of soft skills among employees is a key concern among managers and employers from US-based companies in Asia (as shown through the Milken Institute's surveys) (Klowsen & Lim, 2021), and organizations in Southeast Asia (Mercer, 2021). Creativity is one of the soft skills which are seen as important.

Creativity is important in the uncertain future of work after the COVID-19 pandemic. Creativity and digital competence can contribute to a better economic future as it encourages the generation of new ideas. People who are creative can recognize possible consequences and forecast future scenarios as they consider events from multiple perspectives (Niemi, 2018). Employers' realization of the importance of creativity for the future is highlighted by The World Economic Forum (Zahidi, 2020). In addition, PwC's (2021) annual Global CEO survey indicated that creativity was an important feature in solving the weaknesses in business operating models during the COVID-19 pandemic period. The United Nations (2021) has also stressed on the importance of creativity and innovation in all aspects of human development, especially for developing countries and had marked 2021 as the International Year of the Creative Economy for Sustainable Development. In order to raise awareness of the importance of creativity in human development, 21 April was designated as World Creativity and

DOI: 10.4324/9781003367093-7

Innovation Day. This was because the United Nations Educational, Scientific and Cultural Organization (UNESCO) (2013) identifies creativity and innovation at both the individual and group levels as the true wealth of nations in the 21st century. However, there is an inclusive and sustainable strategy for the development of creativity and innovation among the graduates who are being prepared for the workforce, and in particular, digital creativity.

The need for a sustainable strategy for development of digital creativity and innovation among graduates is important considering that new technologies have transformed the work culture. Employers consider digital investments in the workforce as important for being competitive in the present situation (PwC, 2021). Many of the traditional approaches and work culture in the public and private sectors have been replaced with digital transactions, which benefits the users (Niemi, 2018). Hence, digital creativity and innovation is now an important feature of work as new solutions and ideas need to be developed from multiple perspectives, as technology will continue to evolve and disrupt work scenarios and situations (Niemi, 2018). Hence, the ability to create new knowledge and innovative processes is an advantage and could ensure sustainability in the future (Niemi, 2018).

Digital creativity is not only limited to the creative industry but is required in every field of work. The United Nations (2021) considers creativity to cover all aspects from artistic expression to problem-solving in the context of economic, social and sustainable development. Digital creativity refers to the production-based activities which makes use of digital technologies for manifesting creativity (Hendriksen, Creely, Henderson, & Mishra, 2021). Hence, in the creative economy, organizations in every sector are innovative and producing goods or services which are of value. The creative economy will thrive on the application of skills of creativity and technology for innovation to create products of value, which may be either tangible or intangible (Hearn, 2020). While tangible products may include buildings, facilities and machinery, digital creatives are also non-traditional assets such as art and design, patterns, computer music, digital stories and human–computer interactions (Hendriksen et al., 2021). These non-traditional and intangible assets such as the knowledge, research, design, branding, and software have been the focus of innovation in the developed economies and has been proven to be beneficial as these intangible assets are investments which contribute to the value chain of companies (Hearn, 2020). However, with the digital technologies of IR4.0 such as robotics, artificial intelligence, blockchain, global digital platforms and autonomous systems, there is an urgent need for digital creativity to shape the design, production and consumption of culture and allow for different forms of creative work (Hearn, 2020).

Digital Creativity Capacity Building

Digital creativity has a tremendous potential and digital technologies can be used for the expression of this creativity (Shin, 2010). Hence, in schools and higher education institutions, teachers and instructors should consider methods to develop the creative capacities using digital technologies. There have been studies done on the outcomes and effects of training for creativity among students (Beghetto, 2019). However, less studies are done on the effectiveness of training among teachers in developing students' creative skills (Hafizi & Kamarudin, 2020; Ismail, Yusof, & Pappu, 2016).

Teachers perceive that creativity is not important and do not foster students' creativity (Qian & Clark, 2016). In addition, teacher training programmes do not seem to emphasize the importance and use of digital technologies (Lai Wah & Hashim, 2021). As a result, Malaysian students have low levels of creativity (Jamal, Ibrahim, Abdul Halim & Alias, 2020). On the other hand, teachers who have been trained in productive pedagogies for creativity are able to perform and encourage creative efforts among their students (Amran & Rosli, 2017). Hence, pedagogically sound activities should be designed to foster students' creativity (Didis, Erbas, Cetinkaya, Cakiroglu, & Alacaci, 2016). In this chapter, some of the pedagogical approaches which can be employed are discussed according to an integrated digital competency framework.

Theoretical Framework

Technological pedagogical content knowledge (TPACK) is seen as a necessary digital competence for instructors and teachers in order to cultivate digital creativity. In this section, an integrated digital competency framework is considered for advancing digital creativity.

Technological Pedagogical Content Knowledge Among Instructors and Teachers

TPACK is the knowledge and skills for effectively integrating the content knowledge on the discipline (CK), pedagogical knowledge for teaching (PK) and technological knowledge on the use and application of tools and software (TK) (Mishra & Koehler, 2006). TPACK had changed the concept of creativity in the digital age as it has been shown to be an indicator of creative teaching (Mishra & Koehler 2008). This is because teachers and instructors with TPACK are teaching creatively with technology. Further, the use of digital technologies in education has been shown to develop students' creativity (Henriksen & Mishra, 2015; Mishra & Henriksen, 2018; Reinhold, Hoch, Werner, Richter-Gebert, & Reiss, 2020; Yaniawati, Kariadinata, Sari, Pramiarsih & Mariani, 2020).

However, teachers and instructors seem to lack the pedagogies for effective teaching with technology. Malaysian teachers seem to lack creativity (Abd Samad, Abd Wahab, & Lee, 2016) and the pedagogical skills for creative teaching (Chia & Lin, 2020; Sulaiman, Muniyan, Madhvan, & Ehsan, 2017). Teachers lacked the pedagogical skills to integrate digital technologies in instruction and the methods did not seem to foster students' creativity (Mullet, Willerson, Lamb, & Kettler, 2016; Tee, Samuel, Norjoharudden, Renuka & Hutkemri, 2018; Wan Yusoff, & Che Seman, 2018). The situation was similar in higher education.

TPACK has been used to investigate digital pedagogies for creative teaching (Maor, 2017) and is the model used by the Ministry of Education Malaysia (Mai & Hamzah, 2017). However, when instructors' TPACK at a prominent higher education institution was assessed, instructors had low levels of TPACK and were not sure of the pedagogies for integrating technology although they had higher levels of technology knowledge (Vasodavan, DeWitt & Alias, 2019).

Providing digital skills alone is insufficient for developing the pedagogies for digital creativity among instructors. A holistic training effort for training for TPACK is required where the focus of instruction should not be on the mere transmission of knowledge but on innovative teaching and learning (Dewitt, Alias, & Siraj, 2015). The 21st-century lessons should be designed to incorporate technology and not involve merely the transmission of knowledge, but to incorporate interactive lesson designs to ensure knowledge generation and creativity (Koh, Chai, Benjamin, & Hong, 2015; Vasodavan et al., 2019). The focus of instruction in higher education should be on acquiring skills for interacting, applying, evaluating and creating new knowledge, as well as problem solving (Ronen & Pasher, 2011; Vasodavan et al., 2019). Teacher education programmes could also benefit from an integrative TPACK approach rather than addressing the digital literacies in isolation. Hence, converting the knowledge into skills to develop the capacities among teachers and lecturers, or translational research, is important.

Lecturers need to be technologically and pedagogically competent by having the knowledge and skills to identify suitable tools to teach different content areas rather than just having knowledge on a variety of technologies used in learning environments. The development of human capital that is digitally creative and innovative is needed to face the uncertainties in the future world of work. In addition, this development of human capital needs to start at an early stage which is from the primary school level.

Digital Competencies for the Future

In considering an integrated model of digital literacies with TPACK for teacher education programmes, a digital competency model is needed. Traditional teacher education programmes seemed to have focused on digital literacy models for developing digital skills among teacher trainees (Falloon, 2020). However, a broader digital competency model is needed as in the evolving digital world we live in, where society and culture influence our digital actions (Falloon, 2020; Lund, Furberg, Bakken, & Engelien, 2014; Ottestad, Kelentrić, & Guðmundsdóttir, 2014). This is because there are other elements such as the ethical use of digital technologies, digital citizenship, as well as the health, well-being and safety issues to be considered for building digital competences (Falloon, 2020; Foulger, Graziano, Schmidt-Crawford, & Slykhuis, 2017). More importantly, teacher trainees need to develop a positive attitude towards digital creativity and innovations (Falloon, 2020; Janssen, Stoyanov, Ferrari, Punie, Pannekeet & Sloep, 2013). Hence, a broader digital competency model that includes the diverse knowledge, capabilities and dispositions is needed by future teachers.

Teachers and instructors who were digitally competent would be able to develop graduates with competencies needed for the future. However, the digital competencies required among graduates needs to be defined. The standards for educational technology among students by the International Society for Technology in Education (ISTE) (2021) provides an integrative model of digital competences and encompasses the pedagogies for effective teaching and learning from early childhood, primary and secondary education to work in technical/ vocational training and higher education. These competencies are aligned with UNESCO's Sustainable Development Goals (SDG) in providing access to education, future learning and work, as well as equity and inclusion (ISTE, 2021). As equity and inclusiveness is enabled with digital technologies when online learners are connected and empowered, student-driven processes are enabled and learning can be personalized, engaging and meaningful for students. However, in Malaysia, it is not known to what extend students have access to quality education to prepare them for the future of work. This is even more crucial for students who are in the rural parts of the country. As Malaysian students need to be prepared to be competent to face the future, they will need to have the appropriate digital competences. Hence, the ISTE can be a mechanism to encourage and improve the use digital technologies to develop a digitally skilled workforce who are able to solve problems and are digitally creative.

The standards state that students should be empowered learners, digital citizens, knowledge constructors, innovative designer, computational thinkers, creative communicators and global collaborators. These standards as a digital competency framework are discussed next.

Empowered Learners

Firstly, as an empowered learner, students are active learners who have the autonomy to select resources and materials relevant to them to build their own networks and customize their learning to achieve their goals (ISTE, 2021). The learner should have the autonomy and flexibility to select their learning path. The vast amounts of open resources on the internet have enabled learners to pursue their interests and extend their horizons for new learning opportunities. In addition to open courseware and Massive Open Online Courses (MOOCs), there are repositories of materials on databases, social media (e.g. *YouTube* and *Vimeo*), simulations and gaming applications on virtual worlds (e.g. *SimCity* and *Minecraft*) and in VR platforms (e.g. *SteamVR* and *Occulus*).

Digital Citizens

Next, as digital citizens who live and learn in an interconnected digital world, students need to be responsible users who employ safe, legal and ethical means (ISTE, 2021). One's digital identity and reputation needs to be managed to avoid irrevocable effects of their actions on social media and other networked environments. Hence, the learner should know the effects of their actions, example *trolling* (intentionally posting offensive messages on the internet to get attention, cause trouble or upset someone) on *Facebook* could have a permanent effect and may influence potential employers who may be able to view the offensive posts. The rights of each individual should be respected, and this include the obligations related to sharing of intellectual property. Hence, media that is proprietary belongs to someone and should not be used without permission or compensation.

Knowledge Constructors

Students as a knowledge constructor could critically curate a variety of resources using digital tools produce creative artefacts and make meaningful learning experiences (ISTE, 2021). Learners would need to develop new ideas and theories to solve problems. The knowledge management framework seems to support this approach (DeWitt & Koh, 2020). The information and resources could be compiled on a wiki, a web page or on content curation platforms such as *Nearpod* and *Blendspace* and the ownership of the new knowledge shared is made evident.

Innovative Designer

The student as an innovative designer uses design processes to identify and solve problems (ISTE, 2021). Authentic problems are addressed and theories and ideas to solve these problems are tested. At the same time, innovative artefacts are created to solve the problems. Digital applications are used to plan and manage the design thinking processes and develop prototypes. The innovative designer could be designing three dimensional (3D) objects to investigate the physics of movements and friction, or be designing objects for utility or aesthetic purposes, and evaluating the physical prototype by printing it with a 3D printer. A makerspace culture provides the space for innovation and exploration.

Computational Thinker

As a computational thinker, students employ computational thinking strategies when understanding and solving problems as they formulate problems and define them suited to technology (ISTE, 2021). Data which is compiled is analysed to abstract models and algorithmic thinking for exploring solutions. Hence, projects in which students have the opportunity to collect data and use digital tools to analyse and represent the data in a suitable manner (graphs, pie charts or graphics) among peers. Computational thinking can be developed when tasks which require programming skills are presented. The learner needs to understand the problem, break the problem into components, and create sequence of steps to develop models and create solution to be tested. Digital applications enable the problem and the model to be visualized, and enable the testing processes. Applications which simulate the processes of logical and computational thinking, making use of visual programming to train and develop computational thinking strategies can be utilized in virtual worlds such as *Minecraft* and *CospacesEdu*.

Creative Communicators

Next, students as creative communicators should be able to express themselves creatively and clearly for a variety of purposes using platforms, tools, styles, formats and digital media appropriate to their goal (ISTE, 2021). Communication can be through a variety of digital applications: visualizations through graphics, posters, mindmaps or videos; models which have been developed; or simulations of situations. Emphasis on content which is original and responsibly created and published for the intended audiences.

Global Communicators

Finally, students as global collaborators are able to broaden their perspectives and enrich their learning by collaborating with others and working in teams both locally and globally (ISTE, 2021). Digital tools can optimize the connections with other learners from a variety of cultures and background, and engaging thus engaging with other learners, broadens mutual understanding and learning. Collaborative technologies enable one to work together and examine issues and problems from multiple viewpoints on projects and work towards common goals.

Hence, to support the development of these competencies among students, the ISTE (2021) suggests that educators have the following standards:

i   Educators need to be empowered professionals with technological pedagogical content knowledge and who reflect on practice.
ii   Educators are life-long learners, and staying current with advances in learning sciences.
iii   Educators are leaders who support student empowerment and advocate for equitable access to technology and content.
iv   Educators are responsible digital citizens who contribute to the digital world and online communities with an appropriate digital learning culture.
v   Educators as a learning catalyst who collaborate with both students and colleagues to solve problems and improve practice.
vi   Educators who design authentic experiences for their learners needs in an inclusive environment.
vii   Educators facilitate learning with technology to support students.
viii Educators manage and use learning strategies in digital platforms in virtual environments and hands-on makerspaces.
ix   Educators create learning opportunities for innovative design and computational thinking for problem solving.
x   Educators nurture creativity and creative expression.
xi   Educators are analysts who use data to support students in achieving their learning goals.

However, teacher training programmes may not adhere to these standards. Further, instructors in higher education may also not be familiar with pedagogies for empowering learners. Training for teachers and instructors currently lacks an integrated approach as the tendency is to develop the knowledge of the pedagogies (PK) and the content (CK) separately (Falloon, 2020). In addition, there may also be lack of awareness on the knowledge and skills required to build digitally creative graduates.

Hence, an integrated TPACK approach is needed. Instructors of preservice and in-service teacher training need to firstly make trainees and teachers aware of the importance of digital technologies for developing creativity among students in preparation for the future of work. In addition, teacher training needs to be designed to integrate digital skills with pedagogy in the different content areas to foster digital creativity. The ISTE standards may be a guide for a TPACK approach among teachers and pre-service teachers to develop digital literacy and creativity among students.

Pedagogical Approaches for Fostering Digital Creativity

The digital competences as outlined in the ISTE standards can be considered for a pedagogical approach in fostering digital competences (ISTE, 2021).

Digital Creativity for Empowered Learners

Empowered learners have the autonomy to be flexible in their learning paths, to retrieve information from databases and platforms made accessible, with the aid of search engines. This is in line with knowledge management processes where knowledge acquisition is the first stage of a knowledge creation framework, where the information is made accessible (DeWitt & Koh, 2020; Vasquez-Bravo, Sanchez-Segura, Medina-Dominguez, & de Amescua, 2013). However, the learner needs to critically evaluate the information for its utility before internalizing the knowledge with his own informal, tacit knowledge (Vasquez-Bravo et al., 2013). This is where the instructors play a role in designing a learning environment which is open and flexible, where learners have access to a variety of engaging resources through databases and repositories relevant to learning and evaluate the suitability of resources. Suggestions and opportunities for the learner to pursue their interests and dwell deeper into topics which interest them is enabled with the availability of open courses and MOOCs.

Teachers and instructors should be inclusive by providing access to a variety of materials. Options to use innovative resources with augmented reality, with hotspots that trigger additional information, by providing virtual manipulatives, 3D objects the learner can explore, probe and develop a better understanding of anatomical or mechanical structures, is made possible with VR technologies. At the same time, teachers and instructors could also provide access to virtual environments to be explored, both in and out of this world. In virtual worlds and VR, learners have the autonomy to explore and choose their own path, and interact with the environment. Hence, developing empowered learners means providing a variety of resources to cater to learners' different learning modalities and providing different media (text, image, videos) and different formats (e.g. jpg, mp4 and stl).

In considering VR for education, it needs to be noted that while VR has been used for entertainment and gaming, it is used less for instruction. VR has been used for inclusive education as physically and mentally challenged students can participate in learning (Cheung et al., 2022). In life, VR is used for virtual tourism (e.g. VR apps such as *Within* & *YouVisit*) and real estate viewing (e.g. iProperty and Gamuda Land). In engineering, VR is used to test materials and designs, such as the structural integrity of construction materials for bridges (Li, Badjie, Chiu, & Chen, 2018), and for improving workplace designs (Caputo, Greco, D'Amato, Notaro, & Spada, 2018). In education, VR has been used for training in the automotive industry (Dede, Abdullah, Mulyanti, Rohendi, & Sulaeman, 2020) and to visualize difficult concepts in science and mathematics education, but it has not been used extensively for education in Malaysian schools. This could probably be because teachers are not aware of these technologies and fear that they may encounter problems when using VR for learning. A survey of preservice teachers in Australia also indicated that although they could be highly motivated in using VR themselves, the lack of support and perceived difficulty was a barrier for using VR in the classroom (Bower et al., 2020).

On the other hand, VR has the potential to evoke emotions and engagement, whether used in the form of learning games (e.g. VR apps for learning Biology, *InCell* and *InMind* in Figure 7.1) (Tan, DeWitt & Alias, 2020), or tours for exploring culture and discovering new places (Chan, DeWitt, Rhett, 2020). The immersive nature of VR allows the learner to experience high-fidelity 3D environments (Bower, DeWitt & Lai, 2020). The use of Head Mounted Displays (HMDs) with sophisticated body tracking systems and controller devices for interactions provide for a high-degree autonomy in the learner (see Figure 7.2). This also includes the stand-alone VR headsets either with controllers or without, can also be used for learner autonomy (See Figure 7.3). The advantage of VR is that learners are able to construct their knowledge while in the virtual environment with possibilities of being teleported to places out of this world or being transduced to microworlds such as in the brain (Bower et al., 2020).

Figure 7.1 VR Apps for Learning Biology, *InMind*.

Figure 7.2 Tethered VR with Head Mounted Displays (HMDs), Sophisticated Body Tracking Systems and Controller Devices Connected to a High-End Laptop.

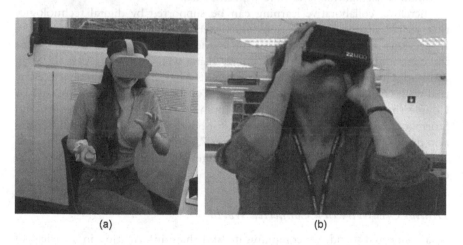

(a)   (b)

Figure 7.3 Standalone VR Headset with Hand controller and VR Headset with Mobile Phone.

However, distractions in the virtual environment as well as lack of familiarization with the interface and other technical problems could affect the effectiveness of learning Bower and Sturman (2015).

Hence, being an empowered learner means the learner is able to personalize his learning and is thus engaged. And, an engaged learner is a creative learner. Empowerment can then enhance digital creativity.

Digital Creativity for Creation of New Knowledge

In order to create new knowledge, knowledge acquired has to be made explicit and shared publicly through visualizations in a formal and systematic manner through either graphics, mental maps or other means (Dalkir, 2011; Vasquez-Bravo et al., 2013). In the knowledge management framework, knowledge is personally articulated and applied with the aid of digital technologies, the aim is to generate new knowledge and this is done creatively. The thinking processes are made explicit for others to scrutinize (DeWitt & Koh, 2020). Hence, with the knowledge now visible to the community, processes such as socialization and argumentation is used to create and verify the new knowledge (Vasquez-Bravo et al., 2013). The interactions with the community enable the explicit knowledge of the community to be developed (Vasquez-Bravo et al., 2013). In this way, there is generation of new knowledge and knowledge creation (DeWitt & Koh, 2020).

A pedagogy which supports this approach is collaborative problem solving, which allows for interactions within a community to create new knowledge. The interactions during collaboration allows for inquiries, debates and justifications to be done so as to achieve work-oriented activities are implemented (Vasodavan et al., 2019). Hence, collaborative learning is enabled when interactions within a community is employed to achieve a specific goal.

Therefore, collaborative learning can be supported by digital technologies that enable sharing, visualizations and interactions (Boza & Conde, 2015). In previous research, opportunities for learners to exhibit their knowledge to the community has been done effectively using wikis (DeWitt, Alias, & Siraj, 2014), discussion forums (Vasodavan, DeWitt, Alias, 2020), and interactive virtual walls (*Padlet*) (DeWitt & Koh, 2020). Collaborative learning apps, such as Coggle (https://coggle.it/) for mindmapping, Miro (https://miro.com), and Stormboard (https://stormboard.com/), for brainstorming are some of the apps also possible for collaborative work (Figures 7.4 and 7.5).

Hence, creating new knowledge encourages digital creativity as knowledge can be externalized and shared.

Digital Creativity in a Connected World

In a connected world, we communicate and share information in a variety of ways. Malaysian learners are digital citizens as they are constantly networking on social media. Digital creativity is being able to connect and share information. Social media has also been a source of content, and for learners to identify and follow experts in their field of expertise. The affordance of social media is due to the publishing, extensive sharing possibilities, and the new forms of interactions possible. Hence, being connected also provides more possibilities for better communication and collaboration. Hence, for learning creatively, a shared purpose should be identified and the contributions and interactions among the group should be encouraged.

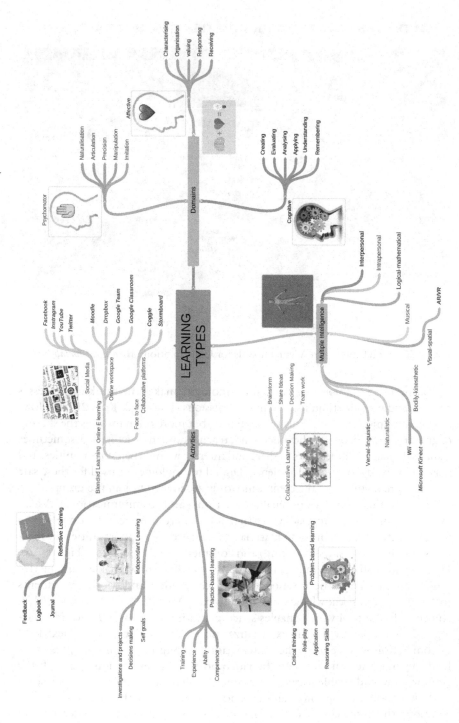

Figure 7.4 A Mindmapping Application Tool, *Coggle*.

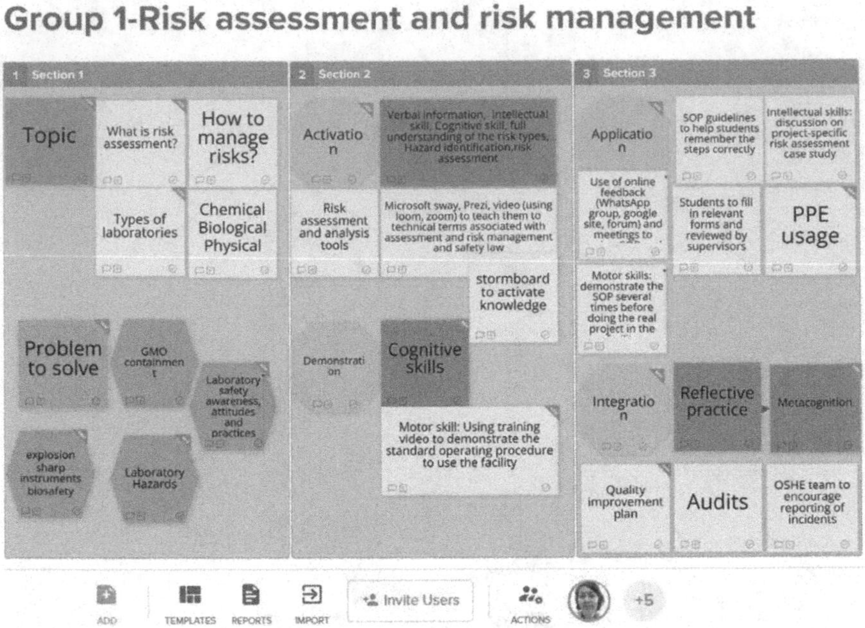

Group 1-Risk assessment and risk management

Figure 7.5 Digital Interactive Virtual Boards for Collaboration: *Miro* and *Stormboard*.

Sharing one's opinions with digital technologies makes the thinking processes explicit, and provide stimulus, enabling discussions (DeWitt & Koh, 2020). However, communication is not only through text, but more often involves the sharing of graphics and videos. On the other hand, teachers and instructors may sometimes want "safe spaces" for learners to share and interact within their communities, and may not be ready for a global audience. Digital technologies be used for these safe spaces where members of the community only are invited to join the groups.

Collaborative applications enable reactions and communications. As an example, *Padlet* is an interactive virtual wall where posts can be done in the form of text, images or links to materials. There are also opportunities for reactions (e.g. likes and stars for voting) and comments to the posts (see Figure 7.6). Hence, internalization and was occurring. Studies show that an interactive *Padlet* could enable the creation of new knowledge and motivate students to learn better as knowledge is externalized (DeWitt, Alias & Siraj, 2015). Another digital tool which enables sharing and interactions is *FlipGrid*. Sharing videos was a way for the externalization of knowledge and the next level of connection was enabled when interactions among the community was enabled through the comments on the video posts. The interactions through the comments would enable learners to evaluate, defend and justify their opinions.

Digital tools can optimize and expand connections with other learners. In addition, the interactions enable the sharing of cultures among learners of

Figure 7.6 Reactions and Comments on an Interactive Virtual Wall.

different background, broaden mutual understanding and examine issues and problems from multiple viewpoints, and work towards common goals.

Digital Creativity for Students as Designers

The design processes where solutions to problems are identified and solutions, designed encourages creativity. In the design thinking approach, problems are reframed in a human-centric way, and more ideas are created through brainstorming, followed by prototyping and testing (Foster, 2019). The design thinking proses have been use for enhancing creativity and developing innovative solutions in many fields (Liedtka, 2018). During the design process, users and consumers needs are considered important and their feedback is constantly needed. Then, new ideas emerging with the brainstorming process are developed and tested. Collaboration during the process helps in idea generation and in developing solutions because ideas are discussed, challenged, defended and new ideas emerge (Liedtka, 2018).

There could be a diverse variety of artefacts in the design process such as a flowchart created for a more efficient process, a storyboard for a new approach in practice, structures with aesthetic value for support in a garden or even a digital device in an energy generator to function in a virtual world. A solution that students could decide to develop would be for communicating awareness among a selected community. Hence, designing posters using a poster-development tool such as Canva, and producing story-boards and videos arousing an emotional response.

The products, whether intermediate such as the story-board, or a final product such as the video, would be constantly evaluated for the student as designer to improve on. In another example, if the process is to solve a problem and design a physical 3D object which does not exist yet, the digital 3D object could be designed using a 3D modelling software (e.g. *Sketchup*), and a plastic version of the object could be produced with a 3D printer. Hence, teachers and instructors should pose problems and create opportunities for students to design prototypes as solutions. Providing a makerspace, either physical or virtual, where solutions can be safely prototyped and tested, would encourage collaboration and creative digital solutions.

Digital Creativity for Computational Thinking

Problems related to technology might require abstract models and algorithmic thinking, which are computational thinking strategies, to explore solutions. Computational thinking is a way of thinking which involves the decomposition of a problem, abstracting the main ideas, algorithmic design, debugging, iteration and generalization (Kong, 2019). The potential of digital media can be enhanced when designing interactions with basic coding. This is how robotics for development of computational thinking for problem solving can be introduced at primary school level as visual coding programming can be used to design interactions with a "robot" (Kong, 2019). Projects where learners collect data, analyse and represent the data in a suitable manner with digital tools, optimize the thinking process. Computational thinking includes programming empowerment as this is an important feature for cultivating creativity and problem-solving skills (Kong, 2019). Hence, learners should be allowed the freedom of designing their own unique solutions to problems, and this can be done in 3D virtual worlds such as *Minecraft*.

Some digital applications enable the problem and the model to be visualized and facilitates the testing processes. An application which simulates the processes of logical and computational thinking and makes use of visual programming for designing interactions in virtual environments is *CospacesEdu* (see Figure 7.7).

Figure 7.7 Visual Programming in *Coblocks* for Designing Interactions in Virtual Environments.

The visual coding tool *Coblocks* develop computational thinking strategies among learners. Hence, animations and movements can be designed as the learner solves problem in design a virtual environment and applies computational thinking strategies.

Virtual worlds can be used for the development of computational thinking. Hence, teachers and instructors should be providing opportunities for learners to work with data, design prototypes and test the prototypes in an environment that can encourage diverse ideas.

Conclusion

Digital creativity is important as it ensures the sustainability of the economy. In order to prepare our graduates with digital creativity for the digital transformation, we need to reskill our teachers and instructors. Firstly, that needs to be awareness that the competency that is required is not just on the technology or the pedagogy, which is trained in isolation, but an integrated TPACK. Administrators and managers of teacher training institutes and institutes of higher learning as well as the Human Resource Department need to be aware of the competences required. In addition, instructor certification of the TPACK and skills should be considered. This is to ensure that instructors involved in training would have the required certification.

Secondly, human capital development needs to focus on competences for developing creativity. Teacher education needs to be reconceptualized as all teachers are digital consumers and users who need to develop students' digital competencies. For this purpose, an integrative digital competency framework that considers pedagogy or TPACK for teacher education. The digital competency framework should empower learners and encourage production-based activities; solving authentic problems and collaborative team work would encourage development of creativity. These competences need to be developed as early as possible, even at the preschool level and to be nurtured throughout an individual's educational journey so as to produce graduates who are digitally creative. Hence, a framework for developing teachers' capacities and creativity which would lead to creative behaviour among their students is required. This is important for developing human capital which is prepared for the digital transformation in Malaysia. In summary, to develop graduates' capacities for creativity, the instructors' digital and pedagogical skills and TPACK, as well as a supportive learning environment is required (Hendriksen et al., 2021).

Digital creativity and TPACK does not seem to be emphasized in preservice teacher education in teacher training institutes. Although the government has emphasized the importance of technology for education since the Smart School concept was launched in 1996 and courses for developing ICT skills have been conducted to develop teachers' capacities (DeWitt & Alias, 2020), training seemed to focus on ICT skills. There does not seem to be sufficient training for developing capacities in pedagogies for the digital era, or on TPACK. Although efforts in offering courses for developing TPACK and integrating technology with learning exist, these are few and sparse. More often, preservice teachers are trained in technology skills separately from the pedagogy and content. The is also an issue in technologically advanced countries like Australia (Falloon, 2020). Teacher training has been critiqued as being ineffective as teachers lack the deep understanding, knowledge and capabilities for educating students to cope with the problems in life and instead focuses on contextually-devoid, isolated technical skills (Falloon, 2020). Hence, what is required is for instructors of preservice and in-service teacher training to make teachers aware of the importance of an integrated framework, TPACK, and to design training to integrate digital skills with the pedagogy in the content areas to foster digital creativity among students. For this purpose, teacher education programmes need to be restructured. Similarly, in higher education institutions, instructors should be provided training and certification in developing TPACK for creative teaching with technology. At the same time, administrators and leaders in the institutions should be made aware of the importance and need of cultivating digital creativity among students.

Pedagogical approaches that empower learners to be flexible in their learning paths, and to enable students to create their own knowledge and design solutions to problems are some of the approaches for enhancing digital creativity. Strategies that employ computational thinking processes would encourage students to be digitally creative and to share their creative products. As students share their products and experiences on a platform which can be accessed by others, learning becomes meaningful (Laurillard, 2012). Further, interactions on a shared platform enables the students' tacit knowledge to be made visible to the community. At the same time, students need to present their product in an attractive manner, and hence develop their digital creativity.

Digital creativity is an important construct that is needed for the future of work and for economic viability as it ensures that that is value in the new ideas that are developed. However, digital creativity is a complex construct which is still widely debated in the literature. Administrators may consider creativity as the use of technology in education while teachers may consider creativity when a tool is integrated in teaching (Hendriksen et al., 2021). On the other hand, students would consider the use of technology creative when they were engaged with the process of learning. In general, these definitions are similar as there is always a value to the new idea or method which is not bounded by conventional thinking processes (Sternberg & Hedlund, 2002). As technology becomes more pervasive in our lives and automation and artificial intelligence becomes important in the economy, digital creativity becomes a necessary competency. Our graduates need to have these competences in order to be competitive in the future. Hence, capacity development using the pedagogical framework for these competences to be cultivated is needed at all levels: in the primary and secondary schools, as well as in higher education.

References

Ab Samad, R. S., Abd Wahab, H., & Lee, Y. N. (2016). The factor of principal instructional leadership roles that contributes the most to teachers' creative pedagogy in Kuala Pilah primary schools, Negeri Sembilan, Malaysia. *Juku: Jurnal Kurikulum & Pengajaran Asia Pasifik*, 4(4): 44–52,

Amran, N., & Rosli, R. (2017). *Teachers' Understanding in 21st Century Skills*. Prosiding Persidangan Antarabangsa Sains Sosial & Kemanusiaan. Persidangan Antarabangsa Sains Sosial & Kemanusiaan, April 2017, Kolej Universiti Islam Antarabangsa. http://conference.kuis.edu.my/pasak2017/

Beghetto, R. (2019). Creativity in Classrooms. In Kaufman, J. C. & Sternberg, R. J. (eds.), *The Cambridge Handbook of Creativity* (pp. 587–606), Cambridge: Cambridge University Press. doi:10.1017/9781316979839.029

Beghetto, R. A. (2019). *Beautiful Risks: Having the Courage to Teach and Learn With Creativity*. Washington DC: Rowman & Littlefield.

Bower, M., DeWitt, D., & Lai, J. W. M. (2020). Reasons associated with preservice teachers' intention to use immersive virtual reality in education. *British Journal of Educational Technology*, 51: 2214–2232. doi:10.1111/bjet.13009

Bower, M., & Sturman, D. (2015). What are the educational affordances of wearable technologies? *Computers & Education*, 88: 343–353. DOI: 10.1016/j.compedu.2015.07.013.

Boza, Á., & Conde, S. (2015). Training, attitude, use and impact of Web 2.0 in higher education: Scale validation/Formación, actitud, uso e impacto de la Web 2.0 en educación superior: Validación de una escala. *Cultura y Educación*, 27(2): 372–406. doi: 10.1080/11356405.2015.1034531

Caputo, F., Greco, A., D'Amato, E., Notaro, I., & Spada, S. (2018). On the use of virtual reality for human-centered workplace design. *Procedia Structural Integrity*, 2970308. AIAS 2017 International Conference on Stress Analysis, AIAS 2017, 6–9 September 2017, Pisa, Italy

Chan, DeWitt & Rhett (2020). *VR4ICC: Increasing Intercultural Awareness*. International Summit of Innovation and Design Exposition (INISIDE) 2020. Universiti Malaya

Cheung, J. C., Ni, M., Tam, A. Y., Chan, T. T., Cheung, A. K., Tsang, O. Y., Yip, C., Lam, W., & Wong, D. W. (2022). Virtual reality based multiple life skill training for intellectual disability: A multicenter randomized controlled trial. *Engineered Regeneration*, 3(2): 121–130. DOI: 10.1016/j.engreg.2022.03.003.

Chia, H. M., & Lim, C. S. (2020). Characterising the pedagogical practices in mathematics lessons among selected Malaysian primary schools. *The Mathematics Enthusiast*, 17(1): 307–323.

Dalkir, K. (2011). *Knowledge Management in Theory and Practice* (2nd ed.). Cambridge, MA: MIT Press.

Dede, Abdullah, A. G., Mulyanti, B., Rohendi, D., & Sulaeman (2020). TVET learning innovation on automotive virtual laboratory based on cloud openstack. *Journal of Technical Education and Training*, 12: 51–60.

DeWitt D., & Alias N. (2020). Computers in Education in Developing Countries, Managerial Issues. Tatnall, A. (ed.), *Encyclopedia of Education and Information Technologies*. New York: Springer, Cham. DOI:10.1007/978-3-030-10576-1_125

DeWitt, D., Alias, N., & Siraj, S. (2014). Wikis for collaborative learning: A case study of knowledge management and satisfaction among teacher trainees in Malaysia. *Procedia - Social and Behavioral Sciences*, 141: 894. DOI: 10.1016/j.sbspro.2014.05.156

DeWitt, D., Alias, N., & Siraj, S. (2015). *Collaborative Learning: Interactive Debates Using Padlet in a Higher Education Institution*. International Educational Technology Conference (IETC 2015).

DeWitt D., & Koh, E. H. Y. (2020). Promoting knowledge management processes through an interactive virtual wall in a postgraduate business finance course. *Journal of Education for Business*, 95(4): 255–262. DOI: 10.1080/08832323.2019.1635977

Didis, M. G., Erbas, A. K., Cetinkaya, B., Cakiroglu, E., & Alacaci, C. (2016). Exploring prospective secondary mathematics teachers' interpretation of student thinking through analysing students' work in modelling. *Mathematics Education Research Journal*, 28(3): 349–378.

Falloon, G. (2020). From digital literacy to digital competence: The teacher digital competency (TDC) framework. *Educational Technology Research & Development*, 68: 2449–2472. DOI: 10.1007/s11423-020-09767-4

Foster, M. K. (2019). Design thinking: A creative approach to problem solving. *Management Teaching Review*, 6(2): 123–140. DOI: 10.1177/2379298119871468

Foulger, T., Graziano, K., Schmidt-Crawford, D., & Slykhuis, D. (2017). Teacher educator digital competencies. *Journal of Technology in Teacher Education*, 25(4): 413–448.

Hafizi, M. H. M., & Kamarudin, N. (2020). Creativity in mathematics: Malaysian perspective. *Universal Journal of Educational Research*, 8(3C): 77–84. DOI: 10.13189/ujer.2020.081609

Hearn, G. (Eds.) (2020). *The Future of Creative Work*. Cheltenham, UK: Edward Elgar Publishing. DOI: 10.4337/9781839101106

Henriksen, D., Creely, E., Henderson, M., & Mishra, P. (2021). Creativity and technology in teaching and learning: A literature review of the uneasy space of implementation. *Educational Technology Research & Development*. DOI: 10.1007/s11423-020-09912-z

Henriksen, D., Hoelting, M., & Deep-Play Research Group (2016). A systems view of creativity in a You-Tube world. *TechTrends*, 60(2): 102–106.

Henriksen, D., & Mishra, P. (2015). We teach who we are: Creativity in the lives and practices of accomplished teachers. *Teachers College Record*, 117(1): 1–46.

International Society for Technology in Education (ISTE) (2021). *The ISTE Standards*. https://www.iste.org/iste-standards

Ismail, Z., Yusof, Y. M., & Pappu, H. (2016). Creativity fostering behavior of mathematics teachers through the implementation of school based assessment. In A. Rogerson (ed), Proceedings of the 12th International Conference of the Mathematics Education for the Future Project: the Future of Mathematics in a Connected World, September 20–26, 2014, Montenegro 2014. Mathematics Education for the Future Project.

Jamal, S. N., Ibrahim, N. H., Abdul Halim, N. D. & Alias, M. I. (2020). A preliminary study on the level of creativity among chemistry students in the district of Melaka Tengah. *Journal of Critical Reviews*, 7(16): 752–761.

Janssen, J., Stoyanov, S., Ferrari, A., Punie, Y., Pannekeet, K., & Sloep, P. (2013). Experts' views on digital competence: Commonalities and differences. *Computers & Education*, 68: 473–481.

Klowsen, K., & Lim, Q. (2021). *Future of Work: Insights for 2021 and Beyond*. Santa Monica, CA: Milken Institute

Koh, J., Chai, C., Benjamin, W., & Hong, H. (2015). Technological pedagogical content knowledge (TPACK) and design thinking: A framework to support ICT lesson design for 21st century learning. *Asia-Pacific Education Researcher*, 24(3): 535–543. http://doi.org/10.1007/s40299-015-0237-2

Kong S.C. (2019). Components and Methods of Evaluating Computational Thinking for Fostering Creative Problem-Solvers in Senior Primary School Education. In Kong S. C., Abelson, H. (eds.), *Computational Thinking Education*. Singapore: Springer. https://doi-org.ezproxy.um.edu.my/10.1007/978-981-13-6528-7_8

Lai Wah, L., & Hashim, H. (2021). Determining pre-service teachers' intention of using technology for teaching English as a second language (ESL). *Sustainability*, 13: 7568. https://doi.org/10.3390/ su13147568

Laurillard, D. (2012). *Teaching as a design science*. New York & Abingdon, Oxon: Routledge.

Li, Y., Badjie, S., Chiu, Y., & Chen, W. (2018). Placing an FRP bridge in Taijiang national park and in virtual reality. *Case Studies in Construction Materials*, 8: 226–237,

Liedtka, J. (2018). Why design thinking works. *Harvard Business Review*. 96(5): 72–79. https://hbr.org/2018/09/why-design-thinking-works

Lund, A., Furberg, A., Bakken, J., & Engelien, K. (2014). What does professional digital competence mean in teacher education? *Nordic Journal of Digital Literacy*, 9(4): 281–299.

Lund, S., Madgavkar, A., Manyika, J., Smit, S., Ellingrud, K., Meaney, M., & Robinson, O. (2021). *The Future of Work After COVID-19*. San Francisco, CA: McKinsey Global Institute

Mai, M. Y., & Hamzah, M. (2017). The development of an assessment instrument of technological pedagogical content knowledge (TPACK) for primary science teachers in Malaysia. *Jurnal Pendidikan Sains & Matematik Malaysia*, 7(1): 93–104

Maor, D. (2017). Using TPACK to develop digital pedagogues: A higher education experience. *Journal Computer in Education*, 4: 71–86. https://doi.org/10.1007/s40692-016-0055-4

Mercer (2021). Win with empathy. Global talent trends 2020–2021: Local companion report. *Marsh & McLennan Companies*. Accessed on 16 June 2022 from https://www.mercer.com/content/dam/mercer/attachments/private/global-talent-trends/2021/gl-2021-gtt-southeast-asia.pdf

Ministry of International Trade and Industry (2019). *Industry 4WRD: National Policy on IR 4.0*. Putrajaya. Ministry of International Trade and Industry.

Mishra, P., & Henriksen, D. (2018). Creativity in Mathematics and Beyond. In *Creativity, Technology & Education: Exploring their Convergence*. Cham: Springer.

Mishra, P., & Koehler, M. J. (2006). Technological pedagogical content knowledge: A new framework for teacher knowledge. *Teachers College Record*, 108 (6): 1017–1054.

Mishra, P., & Koehler, M. J. (2008). *Introducing Technological Pedagogical Content Knowledge*. In Annual meeting of the American Educational Research Association (pp. 1–16).

Mullet, D. R., Willerson, A., Lamb, K. N., & Kettler, T. (2016). Examining teacher perceptions of creativity: A systematic review of the literature. *Thinking Skills and Creativity*, 2: 9–30. DOI: 10.1016/j.tsc.2016.05.001

Niemi, T. (2018). Digital technology as a source of creative organizational resource and service delivery: Building a climate for organizational creativity with deliberative democracy. *Journal of Media Critiques*, 4(14): 241–253. DOI: 10.17349/jmc118218

Ottestad, G., Kelentrić, M., & Guðmundsdóttir, G. (2014). Professional digital competence in teacher education. *Nordic Journal of Digital Literacy*, 9(4): 243–249.

PwC (2021). *24th Annual Global CEO Survey*. PwC's 24th Annual Global CEO Survey: A leadership agenda to take on tomorrow. Accessed on 16 June 2022 from https://www.pwc.com/gx/en/ceo-survey/2021/reports/pwc-24th-global-ceo-survey.pdf

Qian, M., & Clark, K. R. (2016). Game-based learning and 21st century skills: A review of recent research. *Computers in Human Behavior*, 63: 50–58.

Reinhold, F., Hoch, S., Werner, B., Richter-Gebert, J., & Reiss, K. (2020). Learning fractions with and without educational technology: What matters for high-achieving and low-achieving students? *Learning and Instruction*, 65: 101264

Ronen, T., & Pasher, E. (2011). *Complete Guide to Knowledge Management a Strategic Plan to Leverage your Company's Intellectual Capital*. Hoboken, NJ: Wiley.

Schwab, K. (2016). *The Fourth Industrial Revolution*. Geneva: World Economic Forum.

Shin, R. (2010). Taking digital creativity to the art classroom: Mystery box swap. *Art Education*, 63(2): 38–42.

Sternberg, R. J., & Hedlund, J. (2002). Practical intelligence, g, and work psychology. *Human Performance*, 15(1–2): 143–160. DOI: 10.1080/08959285.2002.9668088

Sulaiman, T., Muniyan, V., Madhvan, D., Ehsan, S. D., Persekutuan, W., & Lumpur, K. (2017). Implementation of higher order thinking skills in teaching of science: A case study in Malaysia. *International Research Journal of Education and Sciences (IRJES)*, 1(1): 2550–2158. Accessed from www.masree.info Manuscript

Tan, DeWitt & Alias, 2020. *The use of VRBG for Learning Biology*. International Summit of Innovation and Design Exposition (INISIDE) 2020. Universiti Malaya

Tee, M. Y., Samuel, M., Norjoharuddeen, M. N., Renuka, V. S., & Hutkemri. (2018). Classroom practice and the quality of teaching: Where a nation is going? *Journal of International and Comparative Education*, 7(1): 17–33. https://doi.org/10.14425/jice.2018.7.1.17

United Nations (2021). *World Creativity and Innovation Day*. Accessed form https://www.un.org/en/observances/creativity-and-innovation-day

United Nations Educational, Scientific and Cultural Organization (UNESCO) (2013). *Creative Economy Report: Widening Local Development Pathways*. Paris: United Nations Development Programme (UNDP) & UNESCO. Accessed on 16 June 2022 from http://www.unesco.org/culture/pdf/creative-economy-report-2013-en.pdf

Vasodavan, DeWitt & Alias (2019). TPACK in higher education: Analysis of the collaborative tools used by lecturers. *Jurnal Kurikulum & Pengajaran Asia Pasifik*, 7(1): 9–17

Vasquez-Bravo, D.-M., Sanchez-Segura, M. I., Medina-Dominguez, F., & de Amescua, A. (2013). Knowledge management acquisition improvement by using software engineering elicitation techniques. *Computers in Human Behavior*, 30: 721–730. doi:10.1016/j.chb.2013.09.003

Vinothini Vasodavan, Dorothy DeWitt, Norlidah Alias, Mariani Md Nor. (2020). E-Moderation skills in discussion forums: Patterns of online interactions for knowledge construction. *Journal of Social Sciences and Humanities*, 28(4), 3025–3045. doi: 10.47836/pjssh.28.4.30

Wan Yusoff, W. M. & Che Seman, S. (2018). Teachers' knowledge of higher order thinking and questioning skills: A case study at a primary school in Terengganu, Malaysia. *International Journal of Academic Research in Progressive Education and Development*, 7(2): 45–63. DOI:10.6007/IJARPED/v7-i2/4120

Yaniawati, P., Kariadinata, R., Sari, N. M., Pramiarsih, E. E., & Mariani, M. (2020). Integration of e-learning for mathematics on resource-based learning: increasing mathematical creative thinking and self-confidence. *International Journal of Emerging Technologies in Learning (iJET)*, 15(6): 60–78.

Zahidi, S. (2020). The jobs of tomorrow. *World Economic Forum*. Accessed on 16 June 2022 from https://www.imf.org/external/pubs/ft/fandd/2020/12/pdf/WEF-future-of-jobs-report-2020-zahidi.

8   Online Education in Malaysia

Institutional Change to Stimulate Digital Literacy

Nurliana Kamaruddin, Cecilia Cheong Yin Mei, Wah Yun Low, and Azirah Hashim

Introduction

The boom and rapid expansion of information and communication technology (ICT) since the beginning of the 21st century has made the utilization of the internet and digital tools ubiquitous in all aspects of our lives. Many day-to-day services and processes have shifted online and there is an ever-increasing demand for employees with digital literacy. This has translated into various new social challenges including, not only the lack of a formalized education in developing and enhancing digital literacies in education, but also social divides, new security issues, and even infrastructure requirements. The COVID-19 pandemic has also changed the way we work and learn, forcing an even bigger urgency in transitioning towards a population with high digital literacy. The education sector is one of the sectors that have been fundamentally impacted by the pandemic. The pandemic and the subsequent lockdowns have highlighted Malaysia's unreadiness and inadequacy in meeting the shift to online teaching and learning.

There is still a lack of understanding of what needs to be done within the education sector especially when it comes to improving digital literacy, bridging the digital divide, as well as implementing online teaching and learning. This chapter presents an overview of the challenges that the education sector in general, and in Malaysia in particular, faces due to the sudden need to transition to online teaching and learning arising from the pandemic. Although further research will be necessary in order to obtain concrete data and documentation of the experience of students and teachers on the ground, this chapter is an initial exploration of Malaysia's experience with online education. The research for this chapter was conducted via content analysis of academic literature such as journals and book chapters, working papers and reports produced by think tanks, research institutes and government agencies. It is also supplemented with news and reports from the media.

This chapter will first explore the concept of digital literacy, as well as various motivating factors in making the migration from traditional to online teaching and learning. This is followed by an assessment of the challenges faced in the move to online teaching and learning in Malaysia especially in the COVID-19 era. These challenges include the lack of coherent policymaking both at national

DOI: 10.4324/9781003367093-8

and institute levels, the digital divide that exists, the readiness of teachers and students in making the transition, the concern for special needs education and, finally, security concerns involved. It concludes with an assessment of the prospects for digital literacy and online teaching and learning in Malaysia.

Digitalizing Literacy: Expanding on an Understanding.

Literacy is arguably the bedrock to modern day learning and the education level in any society is closely tied to the level of literacy. Universal primary education where children are provided with the training to read and write is considered a basic right and has become a part of the standard when it comes to measuring the development of any individual, community and even country. As the economy of a country develops and mature, the expected level of literacy needed from the workforce also increases. More often than not, the higher the literacy level of a population, the better developed a country is.

Additionally, in a world that is becoming increasingly information oriented, literacy is also directly connected to social well-being as engaging in daily social life requires a varying degree of literacy such as dealing with essential daily services and its bureaucracy like banks and healthcare providers. All of these are taken into account when trying to understand what literacy means and the expected levels of literacy a country's population has. The UNESCO's definition of literacy is:

> The ability to identify, understand, interpret, create, communicate, compute and use printed and written materials associated with varying contexts. Literacy involves a continuum of learning in enabling individuals to achieve their goals, to develop their knowledge and potential, and to participate fully in their community and wider society.

Similarly, for the Organisation for Economic Co-operation and Development (OECD), this means more than the simple ability to read and write, and the definition of literacy encompasses "the ability to understand and employ printed information in daily activities, at home, at work and in the community – to achieve one's goals, and to develop one's knowledge and potential" (OECD 2000: x). This broader approach to literacy by the OECD covers three "domains" which are prose literacy, document literacy and quantitative literacy which allows them to measure the varying degrees of a country's literacy. The measurement of a country's literacy rate is also important to many international organizations such as the UN, the World Bank and the OECD. This is, because it not only provides a gauge of the overall education level of a country but also reflects the country's ability to upskill its population through educational and employment opportunities UNESCO (2021b).

As we progress into the digital age, literacy has evolved and needs to do so even further. The advent of the internet, followed by the ubiquity of smart devices since the introduction of the first iPhone in 2007, meant that much of our daily services have also shifted online. This has also normalized text-based communication and daily interaction via devices and online platforms. Navigating these new means of information provision requires a whole new set of skills on top of traditional literacy commonly gained in primary schooling. Basic literary and numeracy skills now need to be complemented with digital skills in order to truly improve on a population's literacy level.

Although the concept of acquiring and processing information digitally essentially began with the introduction of the first computer, it was Paul Gilster in 1997 that introduced the term "digital literacy" and he defined digital literacy as "the ability to understand and use information in multiple formats from a wide range of sources when it is presented via computers." This definition introduced before the ubiquitous use of the internet as we understand it today captures the initial understanding of how the world was evolving in its expectation of literacy. A definition of digital literacy by UNESCO (2018: 6) explains it as:

> The ability to access, manage, understand, integrate, communicate, evaluate and create information safely and appropriately through digital technologies for employment, decent jobs and entrepreneurship. It includes competences that are variously referred to as computer literacy, ICT literacy, information literacy and media literacy.

There remains a large section of the population that lacks access to the internet altogether and therefore, unsurprisingly, the first skill needed to cultivate a digitally literate population is knowing how to use devices such as computers and smartphones. That said, the ability to access information is merely the tip of the iceberg when it comes to cultivating digital literacy. Perhaps, even Gilster (1997: 2) himself did not imagine the extent of internet penetration in the 21st century when he wrote the book in 1997, but his statement concerning one of the most important competencies when discussing digital literacy being "the ability to make informed judgements about what you find online" was, and is, an accurate assessment concerning the skill needed to navigate the digital realm. Gilster (1997: 3) also identifies a few more core competencies which include the ability to carry out targeted reading using hypertext and hypermedia, learn how to "assemble knowledge" from the vast information available online, learn how to work within the internet (and using the available Internet tools) as well as developing search skill for online information.

Of course, when discussing digital literacy in today's context, these competencies identified by Gilster (1997) are not quite enough. In an age where everyone is expected to use a device one way or another, there are various other aspects that needs to be considered. For some, these include, not only focusing on the cognitive aspect of digital skills but also the social, emotional, and psychological tools which are necessary to better prepare an individual to navigate, obtain and apply information accessed digitally in a positive manner. Digital literacy now needs to consider not only optimizing a learner or user in their skill sets to access and use digital tool but also to ensure the safety and security of users as well. Consider the fact that the approach to learning within a digital space allows for a multitude of platforms and interactivity. The ability to understand symbols – which we consider reading – is not complemented with various interactive media format which provides information in various new forms and is no longer limited merely to written text.

These contents are also inherently reproducible instantly, can be obtained in a non-sequential and non-linear manner due to hyperlinking and for all thought and purposes, and are limited to the entirety of human knowledge. This presents learners with the challenge of navigating an almost bewildering repository of information that no other generation has ever had access to. Additionally, users are not only content consumers but content creator as well. More importantly, the interactivity of the digital realm means that the interactivity, participation and communication between users globally shape the way information is produced and consumed – meaning any learner or user also needs to develop skills of engagement and communication in the digital realm.

Yoram Eshet-Alkalai (2004) categorizes five different types of literacy necessary within the digital literacy framework. These are "photo-visual literacy; reproduction literacy; branching literacy; information literacy; and socio-emotional literacy" (2004: 94). Each of these types consider not only how a learner engages with the various digital mediums but also the skills necessary to navigate, analyse, and reproduce information. The author also considers the skills of interacting with other users within the digital realm as core skills. This is because the unmoderated, unfiltered and instantaneous nature of interaction online is something that did not need to be taken into consideration in previous modes of literacy.

UNESCO (2018: 87) prepared a "proposed digital literacy competence areas" as part of its Digital Literacy Global Framework (DLGF) project. The competence areas directly related to education includes information and data literacy, communication and collaboration, digital content creation, safety and problem solving. Overall, the migration to online teaching can only be done effectively with a good understanding and enhancement of digital literacy competency. These need to be considered in line with what are the motivating factors to embrace the migration to digital education as well as the status-quo of online teaching and learning in Malaysia.

Motivations for Migration to Digital Education

Digital transformation has a great role to play in the education sector, especially in the era of the Fourth Industrial Revolution (IR4.0) with the rapid growth of online education in recent years. In order to be relevant to the current digital transformation, it is thus necessary for educational institutions to evolve along this technological route. From the educational perspective, digital transformation can be seen in teaching, learning, and organizational or management practices (Fleaca, 2017; Thoring, Rudolph, & Vogl, 2018). Educational institutions ought to explore the opportunities that this digital transformation in education can offer and the significant challenges and difficulties that accompany them too.

Government policies in terms of modernization and bringing local educations institutions to compete with the world need to embrace digital initiatives in education, especially in today's economy and in meeting future demands. Ministries of education around the world have incorporated digitalization plans into the national educational agenda and institutional curriculum. The growing competitive environment in higher education and the need to provide quality online instruction is a matter of long-term survival (Hiltz & Turoff, 2005).

Thus, it is crucial that digital transformation in education is sustainable. Through digital transformation in education, the government can slowly bridge the digital divide between geographical regions, especially those in remote places, provided the digital infrastructure is there. Accessibility and equity to education will give the student population a global experience. However, one needs to address the equity and social justice concerns in learning with technology (Chiu & Lim, 2020). The majority of educational studies related to the pandemic are conducted in mainstream schools or city areas. Furthermore, Chiu, Lin and Lonka (2021) noted that more studies are needed to investigate how to use technology to support minority groups such as students with special education need (inclusive education) and students living in rural areas (digital divide).

Other motivating factor to migrate to digital transformation in education is related to university management and streamlining the administrative and operational processes. For example, data security, data protection and data usage, human resource management, infrastructure, students' registration and admission or enrolment technology, tests and assessments, research governance, academic performances, video conferencing for overseas students and online classes, all these data can be used for informed decision making, to assess and further improve the university performances and for strategic planning.

Natural disaster or other catastrophes such as the current COVID-19 pandemic has resulted in school closures and affecting the education system throughout the world, thus immediately transforming the traditional face-to-face teaching to online teaching and learning (Dhawan 2020). UNESCO (2020a, 2020b) has declared that online learning can help stop the spread of the COVID-19 by avoiding direct interactions between people. UNESCO (2020c) has also provided a list of free educational platforms and resources that can be used for online learning according to the needs of each educational institution, providing social care and interaction during school closures. The COVID-19 pandemic has accelerated the pace at which the world is migrating towards a digital education, and this will be discussed further in the following sections.

The pandemic reveals the urgent need to augment the educational system's technological infrastructure, expand the teachers' pedagogical expertise and the students' learning repertoire (Chiu, Lin & Lonka, 2021). Under the circumstances, researchers and practitioners should carefully reconsider the role of teachers, students, as well as the technological environment for online learning and put ongoing efforts to adequately address the underlying epistemological basis of education (Tsai et al., 2013). During such crises or emergencies, online teaching and learning is no longer an option, but a necessity (Dhawan, 2020).

Online learning is a new social process that is beginning to act as a complete substitute for both distance learning and the traditional face-to-face class (Hiltz & Turoff, 2005).

In the era of the IR4.0, and with the digital technology utilizing augmented and virtual reality, mixed reality, cloud computing, big data, AI, platform technologies, chatbots, gamification, etc. in teaching and learning methodologies, it is apparent that sufficient digital skills are part of a core skillset for the educators and therefore need to be an integral part of curricula and learning environment in universities. There is certainly a paradigm shift in education where this new normal provides the impetus to transform its existing paradigm through technology to better meet student needs. This shift in the entire pedagogical approach is to tackle the new market conditions and adapt to the changing situations (Dhawan, 2020). Tull et al. (2017) noted that natural disasters can stimulate our motivation for the adoption of highly innovative communication technology and e-learning tools.

Different levels of education may also be affected or motivated differently in transitioning to a digital model. For example, digital transformation in higher learning institutions can be viewed from a renewed business model perspective (Betchoo, 2016; Rodrigues, 2017; Gama, 2018) and a marketing aspect incorporated into the system. This migration from traditional to online settings has important implications for the finances of the institution. This is especially true given the substantially lower limits on class sizes that have been widely recommended as being optimal for online courses (Eisenhauer, 2013). Some higher learning institutions may have problems with student enrolment, rising costs, lack of traditional sources of funding, university competitiveness, and perhaps due to disaster or catastrophes, all these may consider migrating to online learning.

Universities are increasingly looking to have additional courses and academic programs online as a means of raising students' enrolment and generating higher tuition revenue, although a large investment is required to adopt these new technologies and educational software. Universities that are not equipping themselves to adapt to this new digital era will be left behind. (PricewaterhouseCoopers, 2018). For educators, they must keep abreast the latest change in technology in teaching and how these can affect students' learning outcomes. Teaching tools based on digital technology can further be more student-centred, creative, flexible, engaging and interactive for students (Henriette, Feki, & Boughzala, 2015), and more innovations in digital teaching and platforms, not just technical innovations, but also academic, curricular, organization and structural innovations (Bond et al., 2018).

One of the advantages of online education is its flexibility for those whose distance from campus, schedules, or other limitations restrict their ability to attend traditional face-to-face classes. The value to the student is the flexibility of being able to integrate education with the demands of work and family (Hiltz & Turoff, 2005). Examples are graduate students who have other family commitments and working professionals with job responsibilities which prevent them from attending face-to-face classes, as well as matured and older populations who may have problems with physical mobility, those staying a distance from the universities, or due to limitations of hostel or expensive rentals.

Easy accessibility to such academic programs is also relevant for such students, as compared to not having a degree at all. Such students may then opt for online classes, either in asynchronous or asynchronous formats. Online or e-learning can play a key role in enabling continuous learning for adult matured students, supporting career paths and employability. Distance learning solutions can provide cost effective and ecological means for ensuring that a growing number of people are not excluded from the workforce due to lack of digital skills (UNESCO (2021a). One can access online learning or e-learning anytime and anywhere (Cojocariu et al., 2014). The demands and expectation of students is something that institutions of higher learning will need to cater to.

Graduates of higher learning institutions would certainly benefit from digital transformation in education. An experience in the digital environment is exciting, enriching and an increase in digital literacy and competencies are in demand in the job market. Today's generation of students grow up with technology, are familiar with digital gadgets and thus are expecting such digital knowledge and digital experiences. Almost every student has a smartphone, iPad, laptops or computers and thus these can be used for innovative teaching and learning, although too much of screen time can be harmful to them. With such digital knowledge and skills, they can be a global digital citizen.

It is, therefore, inevitable that the Malaysian education sector sees similar adoption of digital technologies. For example, in 2011, the Malaysian government made the move to equip primary and secondary schools throughout the country with 4G internet connectivity under the 1BestariNet campaign (Shanmugam, Nur Khairunnisha & Gnanasekaren, 2019). There has also been increased effort to incorporate and promote Massive Open Online Courses (MOOC) by Malaysian universities. Although the effort remains limited, it has been seen as a means to allow wider access to university education. Research by Norliza and Mohamad Sahari points out that Malaysian universities need better support and standards to produce quality MOOC resources (2016).

Challenges Posed by the COVID-19 Pandemic

Although the previous section explores the general motivation and benefit for the digitization of education, the sudden emergence of the COVID-19 pandemic has been the main contributor to the immense changes in teaching and learning globally. UNICEF (2021) reported that about one in seven students have missed more than three-quarters of their in-person learning. This unexpected situation, however, witnessed the importance of digital literacy for one's successful continuance of education and career (Vrana, 2016; Sunarjo, 2021). After the onset of the pandemic traditional face-to-face learning and teaching had been replaced by online modes revealing the extent of preparedness of education providers (Mulroony & Kelly, 2020; Toquero, 2020). It was reported that as of 8 April 2020, about 220 million post-secondary students across 175 countries experienced severe disruptions to their education and this constituted 13 percent of the total number of students affected globally (Norzaini & Doria, 2021).

Malaysia was not spared this abrupt transition. In March 2020, education institutions suspended all face-to-face teaching and learning as well as all meetings and other events. The Malaysian Ministry of Education (MOE) and Ministry of Higher Education (MOHE) had announced that all public and private schools and universities would be conducting teaching and learning online until the end of 2020 and this was further lengthened well into 2021 due to the severity of cases. The Movement Control Order (MCO) in Malaysia severely restricted mobility and this necessitated the change to full-time online education in all education levels from pre-school to higher education and challenges faced by the education institutions were manifold. Campuses were closed to all and international students were forced to remain in their host countries or were sent back as borders were closed for international travel.

Administrators and policy makers had to decide how to ensure that studies would still continue, and the well-being of students ensured given the short and long-term challenges affecting all parties in higher education institutions. For example, revenue for the higher education sector were drastically affected as many international students decided not to continue with their studies. Students who were due to graduate faced difficulties in getting interviews and jobs. While some universities were more equipped to deal with the sudden transition, others struggled with many setbacks faced among lecturers and students.

The prolonged lockdown due to the pandemic has also brought student welfare issues such as mental and financial to the fore, but the challenges focused on here are challenges related to online teaching and learning. Although technology had long been used in varying levels throughout all education institutions in Malaysia, the sudden change to full-time remote learning and use of the learning management systems has been a huge challenge for many. This following section covers several of the major challenges that has affected the education sector in Malaysia when it comes to online teaching and learning. This section looks at several specific challenges that Malaysia (and other countries) have faced in wake of the transition to online teaching and learning.

Policy Implications

As Malaysia has never experienced any kind of disaster or pandemic of this scale in the past, it was not surprising that there were no mitigation plans in place. However, after the initial chaotic phase, a number of policies and guidelines were put into place especially to ensure a smoother progression of all activities during lockdown. Most urgently was how teaching and learning could carry on with minimal disruption if possible. Training sessions also need to be held for academic staff to upskill themselves on online and blended teaching.

In Malaysia, the move to *Pengajaran dan Pembelajaran di Rumah [PdPR]* (Teaching and Learning at Home) began in March 2020 alongside the implementation of the MCO. MOE and MOHE were short pressed to provide a comprehensive approach to online teaching and learning at home, scrambling to balance between providing a teaching and learning manual as well as addressing the various shortcomings in Malaysia's infrastructure. The latter in particular is seen as the utmost urgency as any effort to seamlessly transition to online teaching and learning is likely to fail when students do not have access to electronic devices and reliable internet connection. This issue will be discussed further in the following section. Additionally, the government needed to take into consideration alternative means of providing at-home lessons such as via television and radio (Hawati & Jarud, 2020).

When it comes to higher learning institutes and universities, the change to online also needed to take into account not only teaching and learning for local students but research work as well as internationalization activities. Student mobility came to a halt and had to be replaced with virtual mobility wherever possible. Guidelines and procedures had to be implemented for online assessments and examinations. Concerns of students' accessibility to electronic device and connectivity remains the same at the level of higher education but the independence and flexibility of adult learners did broaden the issues that MOHE needed to look into. These include the encouragement of universities to increase remote learning programmes, open and distance learning (ODL) programmes, international mobility programmes and other types of non-traditional, non-linear approaches such as MOOCs and collaborative online international learning (COIL) courses (Bernama, 2021a).

All of this, coupled with the uncertainty of the COVID-19 pandemic, have, in many cases, created confusion amongst education providers, students and society in general. Throughout 2020 and 2021, the schooling schedule has been adjusted several times to accommodate PdPR (Hana Naz, 2021). There is concern about returning to conventional face-to-face learning and the bigger long-term impact on Malaysia's education sector such as the number of students left behind in the transition process as well as the burnout that students face from an increasing workload and added stress from online classes (Bernama, 2021b; Ramachandran, 2021).

Infrastructure Adequacy and the Digital Divide

Policies and guidelines aside, the major concern in Malaysia during the transition to PdPR and online learning is the inadequate infrastructure for connectivity in Malaysia. This is compounded by the gap between resources in the urban and rural areas. These gaps are also a main concern when it comes to the online teaching and learning capacity in Malaysia where the divide is especially felt between schools and institutes of higher education in the rural and urban parts of the country. Students in the rural areas were not prepared with appropriate and adequate equipment or devices to follow lessons online. In addition, there was a shortage of computers for these students and those from the lower income group (collectively known as B40 group) as it is not cheap to own a computer, let alone to enable the ideal situation of having a computer to a child in every home to attend their classes remotely from home. This situation is further increased in areas where Internet access and electricity connectivity are unavailable. In Malaysia, research by the Ministry of Education reports that about 1.85 million students have no internet access while 90 percent have no personal devices to conduct their online learning (Anuar, 2021).

Unreliable technology has also proven to be a challenge as many learners encountered technical difficulties such as slow internet or no internet coverage, and malfunction of mobile devices or computers, which make online learning challenging. Bandwidth issues would affect the quality of the delivery of knowledge by instructors to students. The delivery is also very much affected by low internet speed and the system's function efficiency. Hence, it is important to ensure that technology fundamentals are in place before instituting e-learning. In addition, dependable technology is necessary for the attainment of ease of navigation and online user-friendliness. Dependable technology must be stressed because learning institutions need to implement a robust and dependable technology to proceed on an education's e-learning journey (Smidt et al., 2020).

Academic and administrative staff have also been badly affected by the digital divide (NurHaiza & NurNadia, 2020). For lecturers, optimal broadband connection is crucial and digital devices for recording lecturers, good lighting, a webcam and microphone are also needed. It is also necessary to have technical support that is usually available at the office for managing technical problems that may arise. The ministries, schools and educational institutes have all realized that investment in infrastructure to support remote and blended learning as well as digital literacy among staff and students is urgently needed in the higher education landscape of today.

In order for e-learning to be executed smoothly and successfully at larger institutions, it is necessary that adequate bandwidth is available. Smidt et al. (2020) who conducted a study on the quality of online and blended learning at a Malaysian University reported that 8 Gbps of broadband entering a large institution may not be sufficient for a smooth and undisrupted access to the online e-learning materials and online classes especially during peak hours of internet usage. A larger bandwidth is needed to form a strong base for internet-enabled learning such as virtual meetings, webcasts, live streaming and online videos viewing. In addition, more wireless access points (WAPs), and high-speed network switches need to be installed campus-wide for both WiFi internet connectivity and wired network access, besides the provision of more WiFi hotspots.

On the other hand, internet downtime may also occur for various other reasons. This is further aggravated by the inability in differentiating between what is essentially a network problem and those which are device related. WiFi coverage may suffer from blind spots and penetration range. Some older laptops and certain mobile devices simply cannot connect to WiFi due to the device or configuration problem. Congestion issues are also contributed by the huge downloads of about 200 Terabyte of data per month and most of these data are from Google, YouTube, Facebook, and Instagram websites. Hence, more speed tests need to be carried out to determine blind spots or low internet speed areas so that the institutions can make full use of the available bandwidth. Even within the more urban areas, there is slow migration from the traditional mode of instructional delivery via chalk and talk to online teaching and learning.

Teachers and Students' Readiness

In terms of student readiness for online learning, Chung et al. (2020) investigated if demographic factors made any difference in their readiness to learn, their online learning experiences and intention to continue using online learning. They found that respondents were generally ready for online learning and that females were found to be more ready than male, degree students were more ready than diploma students while female students and degree students were more satisfied with online learning and have better learning experiences compared to male and diploma students. More than half of the respondents indicated that if given a choice, they would not want to continue with online learning in the future.

As online teaching has become the norm, much work has been necessitated for class preparation, transferring content onto the online platform, rewriting materials and assessments and managing online submission of assignments and exams. As lecturers vary in terms of technical expertise, it is not possible to have everyone move smoothly from face-to-face teaching to online teaching. It can therefore be assumed that some of the delivery of online lessons would have been unsatisfactory. More training sessions need to be organized to enable lecturers to be more effective in delivering online classes. There is also a need to synchronize the online platforms that are used by lecturers to avoid problems for students having to deal with different platforms.

There also seems to be a mismatch in the style of teaching between older generation education providers and the learning styles of younger generation learners who are savvier digitally. In relation to this, there seemed to be a lack of emphasis for lecturers to teach innovatively with the use of new digital technologies. Some teachers and lecturers may lack the pedagogical content knowledge to integrate digital technologies in teaching their respective subject areas. Hence, there is an urgent need to empower educators as well as learners with appropriate digital skills to acquire knowledge and to function well in their professions and future careers. Likewise, the lack of digital and technological skills among students would impact their learning process at the primary right up to the tertiary levels, as well as the students' future job marketability (Santos & Serpa, 2017).

Adjustment for Special Needs Students

Moving to online teaching and learning also presents a whole new set of challenges when it comes to special needs education. At the last 2017 census, Malaysia has about 453,258 persons with disabilities (PWD) who are registered with the Department of Social Welfare (DOSM, 2018). Provision of conventional special needs education already faces its own daunting sets of challenges. The challenges of online teaching and learning are even more augmented for special needs students. What works for students in general education may not work for special education students. New concepts must be broken down into manageable parts, taught in isolation and practised frequently, which will be hard to do in digital learning. Moreover, students with special needs requires an individualized educational plan which is a detailed outline for the special education instruction and services for the special needs student to thrive in school. Each programme is designed to meet the individual student's unique needs.

Most special needs students are not able to use computers for long hours at a time and if the student does not know how to use a computer, getting instruction will be even more difficult (Foster, 2020). Using technology to teach and support students with disabilities will be challenging to special education teachers as they have to first learn how to use the online software in their teaching. Students will also need to be taught how to access and learn from those online materials. As teaching needs to be beneficial to the special needs students, additional gadgets will also be necessary.

Creating lessons that can be disseminated online via computer or smart devices may require technical and software knowledge which the special needs teachers may not possess. Although using technology to teach and support students with disabilities is not entirely new to special education teachers, Dr Jenny Root, associate professor of special education, says that distance and remote learning is outside the typical use of technology, because teachers are not physically present to support and engage students with the technology. Hence, most times teachers resort to "use the same software and routines that their students are familiar with" (Duke, 2021).

Special education teachers will also find student engagement to be a challenge. While in the classroom, teachers can adjust instruction and the lessons based on students' feedback. However, in online and distance learning, it becomes more difficult for teachers to know how and when to adjust learning as both parties are remotely apart (Duke, 2021). It is difficult to replicate the visual supports and provide choices for responding in an online format with extensive support needs for the special needs students. Online teaching and learning hence also becomes more difficult for teachers to pick up these students' cues of understanding, or lack of it, through the computer screen.

Many students usually have a teacher aide with them in the physical classroom to help them navigate through the day. This will not be possible while the student is at home in an online teaching and learning situation. Reports has shown that no amount of alternative learning will replace the amount of quality education like when going to school physically (Abdul-Malik, 2020). As such, both teachers and students with special needs will require assistance from the pedagogical perspectives in terms of human resources, hardware equipment and software materials to ensure some success of a digital and distant education for special needs students.

Security

Another major challenge which also has not received as much attention when it comes to online teaching and learning is the issue of security. As discussed earlier, one of the facets of digital literacy isn't only about optimizing usage of digital devices for learning but also ensuring the learner is equipped with the capacity to use the devices and navigate the online environment in safe and secure manner. Research has also shown that there are security issues which should concern educators including, but not limited to, vulnerability to viruses, phishing and malware attacks, as well as privacy and data leak issues (Joshi, Vinay & Bhaskar, 2020). This issue of concern presented itself in several aspects since the transition to online teaching and learning.

As children as young as preschool are now engaging in online teaching and learning, educators and parents need to be aware of the safety of their students and children when utilizing the various platforms. It's been found that children can (and should) also be made aware of basic security matters such as password privacy and personal data but younger children between the age of 5 and 11 will need direct intervention and instruction by adults to enhance this understanding (Kumar et al., 2017). Of course, this does not preclude the need for parents to be more vigilant in supervising their children's online learning experience.

The extended usage of digital tools and online presence have also made students vulnerable to human predatory behaviour such as harassment, grooming and cyber bullying. Even before the pandemic began, authorities in Malaysia such as the Royal Malaysian Police and the Malaysian Communications and Multimedia Commission (MCMC) are aware of the danger when it comes to children online and initiatives such as *"Klik Dengan Bijak"* (be smart when you click) and the introduction of online safety in textbooks have already been launched (UNICEF, 2019; Vijaindren, 2019). However, it is clear that such initiatives need to be ramped up and this education extended not only to students but education providers and parents. The safety of communication platforms used for online teaching and learning have also been widely debated with cases of outsiders "crashing" class sessions being reported (Aliza, 2021).

The Way Forward

Malaysia has certainly been facing an unprecedented turn of events in dealing with the COVID-19 pandemic. In the education sector, the COVID-19 pandemic has highlighted the country's shortcomings when it comes to migrating to online teaching and learning. However, it has also hastened the transition and brought about better awareness at all levels on what are the issues that policymakers, educators, parents, and students need to be aware of. Although Malaysian schools have slowly been reopening since October 2021, there is no doubt that the education landscape in Malaysia has been altered. Improving the infrastructure and connectivity nationwide must be continued and access to devices needs to be considered a need rather than a luxury. More than that, it is also a changing point for the entire education sector.

Online education in not simply a change in the teaching and learning methods but requires a paradigm shift in the whole teaching and learning experience. This means re-evaluating what we mean by literacy and the tools and skills needed to produce a digitally literate society that we need to maximize our human talent. Considering the understanding of digital literacy as well as the motivating factors of the shift to online teaching and learning, it is clear that Malaysia is still mired with various challenges in moving forward with online education. What is feared is that the current generation of students experiencing a period of unprecedented change could become a "lost generation" due to the ineffectiveness of online teaching and learning despite the ministry's claim of "moderate effectiveness" of PdPR (Bernama, 2021c; NSTP, 2021; UNICEF, 2020).

Going forward, it would be necessary for policy makers and education providers to take a deep dive into ensuring that online teaching and learning is improved. It also means setting in place the needed infrastructure and transitionary preparation to continue enhancing the digital literacy and online teaching and learning experience of students and learners nationwide even if, and when, the COVID-19 pandemic subsides. Malaysia needs to embrace the migration to online teaching and learning. In order to do so, there are several key considerations that the government and education providers should take into consideration.

Without a doubt, the priority is the improvement of infrastructure and access. The internet and digital devices should be seen as a necessity rather than a luxury as almost all form of education will be utilizing online teaching and learning sooner or later. There is increased likelihood that all jobs will soon require some form of digital literacy and Malaysia needs to make the investment into supporting its education sector through improving initiatives such as providing laptops for students in the B40 group.

There needs to be increased emphasis on the training and upskilling for teachers and lecturers. Many teachers and lecturers are still confined by traditional teaching methods and struggle with the transition during the pandemic. It is necessary to ensure that our educators are prepared to handle online teaching and learning. Although many are still hopeful that sooner or later students will be able to return to school, it is naïve to assume that conventional classroom learning will be adequate.

Additionally, the effort to revise and evolve the national education curriculum to include online and hybrid learning have to be given priority. New education syllabus and materials need to be produced in a manner that incorporate the means of digital learning. The government's approach needs to be holistic, and this also means taking into consideration those most likely to be left behind in the transition to digital based education. Poorer students with lack of access and devices as well as special needs education needs to be given renewed attention.

The new normal should not be considered the new normal "for now" but to accept that this change need to be long-term and permanent. The understanding of literacy has to shift in order for the Malaysian education sector to meet the challenges ahead. This also means that better research concerning the experience of students and practitioners should be conducted in order to get a clearer picture of what improvements need to be made. As a country that is trying to reach high income status, Malaysia needs an education sector that is able to support the necessary human resource development. In the long run, investing in the education sector is inevitable should the nation wish to secure its capacity to develop for the foreseeable future.

References

Abdul-Malik, J. (2020). *Feeling Forgotten: Students With Special Needs Face Unique Challenges With Virtual Learning.* GPB. https://www.gpb.org/news/2020/08/04/feeling-forgotten-students-special-needs-face-unique-challenges-virtual-learning

Aliza Shah. (2021, October 17). Polis buka kertas siasatan kes PdPR diceroboh orang luar [Police open investigation case on invasion of home teaching and learning session]. *MStar.* https://www.mstar.com.my/lokal/semasa/2021/10/17/polis-buka-kertas-siasatan-kes-pdpr-diceroboh-orang-luar

Anuar Ahmad. (2021, July 16). Kaji "kesilapan" PdPR elak terus bebankan guru, murid [Look into the "mistakes" of home teaching and learning to prevent burdening teachers and students]. *BH Online.* https://www.bharian.com.my/rencana/minda-pembaca/2021/07/839959/kaji-kesilapan-pdpr-elak-terus-bebankan-guru-murid

Bernama. (2021a). Academia day 2021 focuses on empowering nation's higher education digitalisation agenda. *Edge Markets.* Retrieved October 28, 2021, from https://www.theedgemarkets.com/article/academia-day-2021-focuses-empowering-nations-higher-education-digitalisation-agenda

Bernama. (2021b, September). Education ministry: Effectiveness of PdPR method at moderate level. *New Straits Times.* https://www.nst.com.my/news/nation/2021/09/729110/education-ministry-effectiveness-pdpr-method-moderate-level

Bernama. (2021c). Education ministry: Effectiveness of PdPR method at moderate level. *New Straits Times.* https://www.nst.com.my/news/nation/2021/09/729110/education-ministry-effectiveness-pdpr-method-moderate-level

Bernama. (2021, October 5). Academia Day 2021 focuses on empowering nation's higher education digitalisation agenda. *Edge Markets.* https://www.theedgemarkets.com/article/academia-day-2021-focuses-empowering-nations-higher-education-digitalisation-agenda

Betchoo, N. K. (2016). Digital transformation and its impact on human resource management: A case analysis of two unrelated businesses in the Mauritian public service. *2016 IEEE International Conference on Emerging Technologies and Innovative Business Practices for the Transformation of Societies.*

Bond, M., Marín, V. I., Dolch, C., Bedenlier, S., & Zawacki-Richter, O. (2018). Digital transformation in German higher education: Student and teacher perceptions and usage of digital media. *International Journal of Education Technology in Higher Education, 15*(48), 1–20.

Chiu, T. K. F., & Lim, C. P. (2020). Strategic use of technology for inclusive education in Hong Kong: A content-level perspective. *ECNU Review of Education, 3*(4), 715–734.

Chiu, T. K. F., Lin, T.-J., & Lonka, K. (2021). Motivating online learning: The challenges of COVID-19 and beyond. *The Asia-Pacific Education Researcher Volume, 30*, 187–190.

Chung, E., Subramaniam, G., & Dass, L. C. (2020). Online learning readiness among university students in Malaysia amidst Covid-19. *Asian Journal of University Education, 16*(2), 46–58.

Cojocariu, V.-M., Lazar, I., Nedeff, V., & Lazar, G. (2014). SWOT analysis of e-learning educational services from the perspective of their beneficiaries. *Procedia - Social and Behavioral Sciences, 116*, 1999–2003.

Department of Statistics Malaysia (DOSM). (2018). Social Statistics Bulletin Malaysia 2018. *Social Indicators.* https://www.dosm.gov.my/v1/index.php?r=column/ctheme-ByCat&cat=152&bul_id=NU5hZTRkOSs0RVZwRytTRE5zSitLUT09&menu_id=U3VPMldoYUxzVzFaYmNkWXZteGduZz09

Dhawan, S. (2020). Online learning: A panacea in the time of COVID-19 crisis. *Journal of Educational Technology Systems, 49*(1), 5–22.

Duke, J. (2021). *Addressing Challenges in Online Special Education.* Florida State University. https://education.fsu.edu/addressing-challenges-online-special-education

Eisenhauer, J. G. (2013). Student migration to online education: An economic model. *Journal of Academic Administration in Higher Education, 9*(1), 19–28.

Eshet-Alkalai, Y. (2004). Digital literacy: A conceptual framework for survival skills in the digital era. *Journal of Educational Multimedia and Hypermedia, 13*(1), 93–106.

Fleacă, E. (2017). Embedding digital teaching and learning practices in the modernization of higher education institutions. *International Multidisciplinary Scientific GeoConference, SGEM.* https://doi.org/10.5593/sgem2017/54

Foster, L. (2020). *Challenges of Online Learning for Students with Special Needs.* The 21st Show. https://will.illinois.edu/21stshow/story/challenges-of-online-learning-for-students-with-special-needs

Gama, J. A. P. (2018). Intelligent educational dual architecture for university digital transformation. *2018 IEEE Frontiers in Education Conference (FIE)*, 1–9.

Gilster, P. (1997). *Digital Literacy.* New York: Wiley Computer Publications.

Hana Naz Harun. (2021). Online learning to continue, for now. *New Straits Times.* https://www.nst.com.my/news/government-public-policy/2021/06/696357/online-learning-continue-now-nsttv

Hawati Abdul Hamid, & Jarud Romadan Khalidi. (2020). *Covid-19 and Unequal Learning* (33/20; Views). http://www.krinstitute.org/assets/contentMS/img/template/editor/20200426_Covid_Education_v3.pdf

Henriette, E., Feki, M., & Boughzala, I. (2015). The shape of digital transformation: A systematic literature review. *Information Systems in a Changing Economy and Society: MCIS2015 Proceeding*, 431–443. https://aisel.aisnet.org/mcis2015/10

Hiltz, S. R., & Turoff, M. (2005, October). Education goes digital: The evolution of online learning and the evolution of higher education. *Communications of the ACM, 48*(10), 59–64. https://doi.org/10.1145/1089107.1089139

Joshi, A., Vinay, M., & Bhaskar, P. (2020). *Impact of Coronavirus Pandemic on the Indian Education Sector: Perspectives of Teachers on Online Teaching and Assessments* (No. 18; 2). https://www.emerald.com/insight/content/doi/10.1108/ITSE-06-2020-0087/full/pdf?title=impact-of-coronavirus-pandemic-on-the-indian-education-sector-perspectives-of-teachers-on-online-teaching-and-assessments

Kumar, P., Naik, S. M., Devkar, U. R., Chetty, M., Clegg, T. L., & Vitak, J. (2017). "No Telling Passcodes Out Because They're Private": Understanding children's mental models of privacy and security online. In C. Lampe, J. Nichols, K. Karahalios, G. Fitzpatrick, U. Lee, A. Monroy-Hernandez & W. Stuerzlinger (Eds.), *ACM on Human-Computer Interaction* (pp. 1–21). ACM Digital Library. https://doi.org/10.1145/3134699

Mulrooney, H. M., & Kelly, A. F. (2020). COVID-19 and the move to online teaching: Impact on perceptions of belonging in staff and students in a UK widening participation university. *Journal of Applied Learning & Teaching, 3*(2), 17–30. https://doi.org/10.37074/jalt.2020.3.2.15

New Straits Times Press (NSTP) (2021, January 22). NST leader: The "lost generation." *New Straits Times.* https://www.nst.com.my/opinion/leaders/2021/01/659522/nst-leader-lost-generation

Norliza Ghazali, & Mohamad Sahari Nordin. (2016). The perception of university lecturers of teaching and learning in massive open online courses (MOOCS). *Journal of Personalized Learning, 2*(1), 52–57.

Norzaini A., & Doria A. (2021). A critical analysis of Malaysian higher education institutions' response towards COVID-19: sustaining academic program delivery. *Journal of Sustainability Science and Management, 16*(1), 70–96.

NurHaiza, N., & NurNaddia, N. (2020). Impact of pandemic COVID-19 to the online learning: Case of higher education institution in Malaysia. *Universal Journal of Educational Research, 8*(12A), 7607–7615. https://doi.org/10.13189/ujer.2020.082546

OECD (2000). *Literacy in the Information Age: Final Report of the International Adult Literacy Survey.* https://www.oecd.org/education/skills-beyond-school/41529765.pdf

PricewaterhouseCoopers LLP (2018). *The 2018 Digital University: Staying Relevant in the Digital Age.* https://www.pwc.co.uk/assets/pdf/the-2018-digital-university-staying-relevant-in-the-digital-age.pdf

Ramachandran, J. (2021, July 12). UM student union pleads For MOHE To take burnout seriously after death of 2 students. *Says.Com.* https://says.com/my/news/um-student-union-pleads-for-mohe-to-take-the-issue-of-burnout-among-students-seriously

Rodrigues, L. S. (2017). Challenges of digital transformation in higher education institutions: A brief discussion. *Proceedings of the 30th International Business Information Management Association Conference, IBIMA 2017—Vision 2020: Sustainable Economic Development, Innovation Management, and Global Growth.* https://www.scopus.com/inward/record.uri?eid=2-s2.0-85048618825&partnerID=40&md5=65525232d18dbc0ae37a733eb45b100d

Santos, A. I., & Serpa, S. (2017). The importance of promoting digital literacy in higher education. *International Journal of Social Science Studies, 5*(6), 90–93. https://doi.org/10.11114/ijsss.v5i6.2330

Shanmugam, K., Nur Khairunnisha Zainal, & Gnanasekaren, C. (2019). Technology foresight in the virtual learning environment in Malaysia. *International Conference on Computer Vision and Machine Learning.* https://iopscience.iop.org/article/10.1088/1742-6596/1228/1/012068/pdf

Smidt, E., Cheong, C. Y. M., & Kochem, T. (2020). The meaning of quality in online/blended learning at a Malaysian university. *International Journal on E-Learning, 19*(3), 243–263.

Sunarjo, E. (2021). *Why Digital and Information Literacy Are so Important to Education?* Emerald Publishing. https://www.emeraldgrouppublishing.com/news-and-press-releases/why-digital-and-information-literacy-are-so-important-education

Thoring, A., Rudolph, D., & Vogl, R. (2018). The digital transformation of teaching in higher education from an academic's point of view: An explorative study. *The Applications of Evolutionary Computation,* Vol. 10924, 294–309.

Toquero, C. M. (2020). Challenges and opportunities for higher education amid the COVID-19 Pandemic: The Philippine context. *Pedagogical Research, 5*(4), 1–5.

Tsai, C.-C., Chai, C. S., Wong, B. K. S., Hong, H.-Y., & Tan, S. C. (2013). Positioning design epistemology and its applications in education technology. *Educational Technology & Society, 16*(2), 81–90.

Tull, S., Dabner, N., & Ayebi-Arthur, K. (2017). Social media and E-learning in response to seismic events: Resilient practices. *Journal of Open, Flexible and Distance Learning, 21*(1), 63–76.

UNESCO (2018). *A Global Framework of Reference on Digital Literacy Skills Indicators 4.4.2* (No. 51; UIS/2018/ICT/IP/51). http://uis.unesco.org/sites/default/files/documents/ip51-global-framework-reference-digital-literacy-skills-2018-en.pdf

UNESCO (2020a). *COVID-19 Educational Disruption and Response.* https://en.unesco.org/news/covid-19-educational-disruption-and-response

UNESCO (2020b). *Distance Learning Strategies in Response to COVID-19 School Closures* (Issue note No 2.1; Education Sector Issue Notes). https://unesdoc.unesco.org/ark:/48223/pf0000373305/PDF/373305eng.pdf.multi

UNESCO (2020c). *COVID-19 response – Remote learning strategy: remote learning strategy as a key element in ensuring continued learning* (programme and meeting document ED/PLS/2020/01/REV.2). https://unesdoc.unesco.org/ark:/48223/pf0000373764

UNESCO (2021a). *Distance Learning Solutions.* https://en.unesco.org/covid19/educationresponse/solutions

UNESCO Institute for Statistics (2021b). *Literacy Rate.* Glossary. http://uis.unesco.org/en/glossary-term/literacy-rate

UNICEF (2019). Working together for better online child protection. *Press Release.* https://www.unicef.org/malaysia/press-releases/working-together-better-online-child-protection

UNICEF (2020). UNICEF calls for averting a lost generation as COVID-19 threatens to cause irreversible harm to children's education, nutrition and well-being. *Press Release.* https://www.unicef.org/eap/press-releases/unicef-calls-averting-lost-generation-covid-19-threatens-cause-irreversible-harm

UNICEF (2021). Schools still closed for nearly 77 million students 18 months into pandemic – UNICEF. *Press Release.* https://www.unicef.org/eap/press-releases/schools-still-closed-for-77-million-children-18-months-into-pandemic

Vijaindren, A. (2019, December 13). Online child grooming' to be included in textbook next year. *New Straits Times.* https://www.nst.com.my/news/nation/2019/12/547493/online-child-grooming-be-included-textbook-next-year-nsttv

Vrana, R. (2016). Digital literacy as a boost factor in employability of students. In S. Kurbanoğlu et al. (Eds.), *Information Literacy: Key to an Inclusive Society. ECIL 2016. Communications in Computer and Information Science,* Vol. 676, 169–178.

9 E-commerce Expansion in Malaysia

Tham Siew Yean and Andrew Kam Jia Yi

Introduction[1]

The globalisation of internet and the increasing use of digital technologies have transformed the domestic economy, as well as the nature of trade. A digital economy has emerged in Malaysia, while the scope, scale and speed of cross-border trade have also changed with digitalisation. With the internet, cross-border trade can now be conducted with a mere click of the mouse, thereby reducing transaction costs and some entry barriers, such as the fixed costs of having to set up stores in foreign lands. In particular, with the internet, small and medium enterprises (SMEs) now have opportunities to participate in cross-border trade, without having to incur the cost of establishing a physical presence, as well as opportunities to increase the scalability of their operations. Digital technologies can also be used to enhance productivity and the capacity to compete regionally and globally.

Consequently, all countries are rushing to introduce digital technologies, including on trade to improve their respective economic growth prospects. In particular, the emergence of the coronavirus (COVID-19) pandemic in 2020 hastened the rush to embrace digital technologies and digital trade (DT) to overcome the restrictions from prolonged lock-downs in many of these countries. Likewise, in Malaysia, there is also a race to embrace digital technologies as a new driver of growth. The roots of Malaysia's interest in e-commerce can be traced back to the information, communication and technology (ICT) plans in the country. The Seventh Malaysia Plan (Malaysia, 1996) prioritised the development of ICT infrastructure as a means for shifting the Malaysian economy from a production-based (P-based) to a knowledge-based economy (K-based). E-commerce has later been taken into account in all subsequent five-year Malaysia Plans.

In 2016, Malaysia launched the National E-commerce Strategic Roadmap (NESR 1.0) in a bid to embrace e-commerce as a driver of growth. It has six focus areas, namely, accelerating seller adoption of e-commerce, increasing the adoption of e-procurement by businesses, lifting non-tariff barriers, realigning existing economic incentives, making strategic investments in select e-commerce players and promoting national brand to boost cross-border e-commerce. The government subsequently launched the NESR 2.0 in 2021 to further intensify e-commerce adoption and growth; enhance ecosystem development; and strengthen policy and regulatory environment (Digital News Asia, 2021). NESR 2.0 is to be operationalised from 2021 to 2025.

The government seeks to increase the contribution of e-commerce to gross domestic product (GDP) and to assist SMEs' participation in leading international market places for exports. Incentives, such as the e-trade incentives, are offered by Malaysia External Trade Development Corporation (MATRADE) to facilitate SMEs on-boarding e-commerce. The NESR 2.0 aims to onboard 875,000 MSMEs, boost 84,000 of them to adopt e-commerce export and increase the average revenue per user (ARPU) to RM9,500 monthly (Hew, 2021).

The Digital Free Trade Zone (DFTZ) launched in 2017 in partnership with Alibaba also aims to connect Malaysia's SMEs with other players along China's Belt and Road Initiative that will eventually form a Digital Silk Road (Tham, 2020). The logistics arm of Alibaba, Cainiao Network, in an equity partnership with Malaysia Airport Holdings Bhd, has established an ASEAN e-commerce hub at the DFTZ that has stared just operations in 2020 (Gomez et al., 2020).

In view of these developments, this chapter aims to examine the nature of e-commerce in Malaysia, including cross border e-commerce and its drivers as well as to identify the main challenges in moving forward.

Digital Trade and E-commerce: Concepts and Measurements

Despite the growing importance and interest in the digital economy, there is no single recognised and accepted definition of DT. The Organisation for Economic Co-operation and Development (OECD) defines it to be all "digitally-enabled transactions of trade in goods and services, which can be either digitally or physically delivered and that involve consumers, firms and governments" (OECD undated). E-commerce, on the other hand, is defined as the sale or purchase of goods or services, conducted over computer networks by methods specifically designed for the purpose of receiving or placing orders (OECD, 2011). Although some organisations do use DT and e-commerce interchangeably, OECD (2011) maintains a distinction between the two concepts: whether a commercial transaction qualifies as e-commerce is determined by the ordering method rather than the characteristics of the product purchased, the parties involved, the mode of payment or the delivery channel.

Others (e.g. Herbert Smith Freehills, 2018) also distinguish between the two concepts as DT is deemed to encompass more than the buying and selling of goods and services online because it includes cross-border transmission of information and data. For example, it also includes the use of digital technologies such as digitised customs for customs facilitation. Thus, DT does not only include end-products like downloaded movies and video games. It can also include digital services that enhance the productivity and overall competitiveness of an economy, such as information streams needed by manufacturers to manage global operations; communication channels (email and voice over Internet protocol [VoIP]); as well as financial data and transactions for online purchases or electronic banking. E-commerce is viewed as a part of DT.

Unfortunately, even when the OECD definition/framework is adopted, the measurement of DT is fraught with difficulties. For example in the United Kingdom, existing data on national statistics data-banks are based on surveys that were not designed to capture DT, especially cross-border DT activities (Cambridge Econometrics, 2020). Hence, some of digital transactions are not captured in existing data-banks, especially transactions that have cross-border data flows. Additionally, the value of e-commerce transactions and the distinction between domestic and cross-border dimensions are not covered in business and household surveys. Only *potentially* digitally delivered or ICT-enabled services are estimated from the existing Extended Balance of Payments Services (EBOPS) statistics. Experimental statistics on the UK services by modes of supply have used services delivered by Mode 1 (remote delivery) to be equivalent to digital delivery but trade by modes of delivery is not readily available for every country, especially developing countries. There is no international data-bank for DT and e-commerce data. The latter data are generally fragmented and estimated by payment and platform providers, or postal orders which only capture for parcel deliveries. Since there are no international databanks, this study will use country-level data to explore the development of DT and/or e-commerce, based on the availability of data.

Malaysia's Digital Economy and Nature of E-commerce

As in most countries, Malaysia does not collect data on DT. Instead, it collects data on the digital economy, based on the ICT Satellite Account. ICT is defined to include manufacturing, ICT trade, ICT services and content and media. E-commerce data follow the OECD 2015 definition,[2] and it is collected based on Usage of ICT and E-Commerce by Establishment Survey (ICTEC) and the ICT Use and Access by individuals and Household survey (ICTHS) (Hasnah Mat and Mazreha Ya'akub, 2019). For this reason, the analysis of the development of DT in Malaysia can only focus on the development of e-commerce.

Overall, the contribution of ICT to GDP has grown from 16.5 percent in 2010 to 19.1 percent in 2019 (Figure 9.1). The share of e-commerce in non-ICT industries has grown steadily from 3.6 percent in 2010 to 6.2 percent unlike the share of ICT which fluctuated over time. In 2019, its share is the same as in 2010 (12.9percent).

Malaysia's e-commerce income grew from RM398.2 billion in 2015 to RM675.4 billion in 2019. It is mostly domestic (87percent) in origin, with only 12.4 percent contributed by international sources in 2019, which does not distinguish between export and imports (Figure 9.2). It is also largest in manufacturing, followed by services. Although manufacturing dominates domestic income, international income in services is larger than manufacturing for all the years shown (Table 9.1).

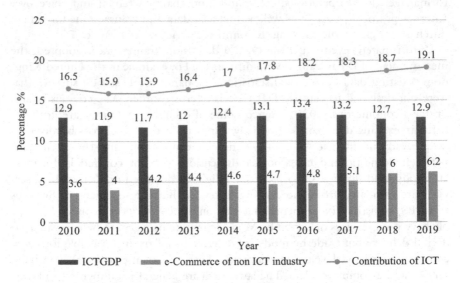

Figure 9.1 Contribution of ICT to the Economy, 2010–2019 (percent).

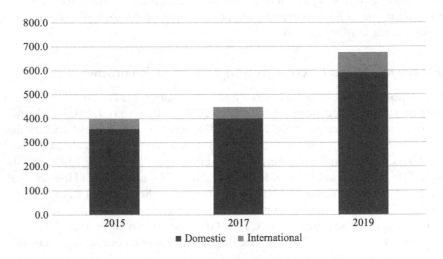

Figure 9.2 E-commerce Income, 2015–2019 (billion).

E-commerce is dominated mainly by B2B, followed by B2C and B2G (Figure 9.3). Manufacturing is dominant in B2B while services are dominated by B2C with a higher share of B2G compared to other sectors.

The imposition of movement control order (MCO) to control the COVID-19 pandemic in 2020 and work-from-home (WFH) restrictions increased the demand for e-commerce activities. According to Mazreha and Norazlin (2021), e-commerce income grew by 22.8 percent annually from RM675.4 billion in

Table 9.1 E-commerce Income (Domestic and International), by sectors, 2015–2019 (RM billion)

(RM Billion)	Year	Total	Domestic	International
Total	2015	398.2	356.8	41.3
	2017	447.8	399.8	48.0
	2019	675.3	591.8	83.5
Agriculture, mining & quarrying, construction	2015	7.1	5.3	2.0
	2017	7.8	5.7	2.0
	2019	10.5	8.3	2.3
Manufacturing	2015	275.9	255.7	20.2
	2017	287.5	266.1	21.4
	2019	354.3	323.7	30.6
Services	2015	115.1	95.9	19.2
	2017	152.6	128.0	24.6
	2019	310.5	259.9	50.7

Source: DOSM.

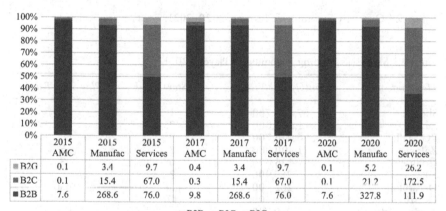

	2015 AMC	2015 Manufac	2015 Services	2017 AMC	2017 Manufac	2017 Services	2020 AMC	2020 Manufac	2020 Services
■ B2G	0.1	3.4	9.7	0.4	3.4	9.7	0.1	5.2	26.2
■ B2C	0.1	15.4	67.0	0.3	15.4	67.0	0.1	21.2	172.5
■ B2B	7.6	268.6	76.0	9.8	268.6	76.0	7.6	327.8	111.9

■ B2B ■ B2C ■ B2G

Figure 9.3 E-commerce Income by Types of Customers (RM Billion), 2015–2017.

2019 to RM896.4 billion in 2020. Thus, the share of e-commerce in Malaysia's ICT economy is expected to increase in 2020 and beyond, as Figure 9.4 shows that quarterly increase in e-commerce revenue since 2020. This is due to digital consumption habits that have increased and developed over the COVID-19 pandemic period, and which is expected to continue as a new way of life (Google, Temasek, Bain&Co, 2021).

According to IPrice, Malaysia's top ten e-commerce merchants are Shopee, Lazada, PG Mall, Zalora, GoShop, eBay, Decathlon, Qoo10, Applecrumby and Sephora (Table 9.2). Social and online media like Twitter, Instagram and Facebook are also popularly used for social-commerce, especially among the young.

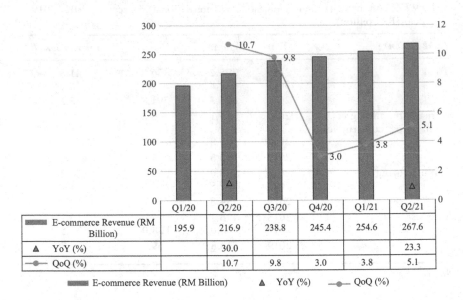

	Q1/20	Q2/20	Q3/20	Q4/20	Q1/21	Q2/21
E-commerce Revenue (RM Billion)	195.9	216.9	238.8	245.4	254.6	267.6
▲ YoY (%)		30.0				23.3
QoQ (%)		10.7	9.8	3.0	3.8	5.1

E-commerce Revenue (RM Billion)   ▲ YoY (%)   QoQ (%)

Figure 9.4 Quarterly E-commerce Income, 2020–2021.

Table 9.2 Malaysia's Leading E-commerce Players, 2021

Merchant	Monthly Web Visits	App Store Rank	Play Store Rank
Shopee	57,566,700	1	1
Lazada	14,290,000	2	2
PG Mall	6,976,700	9	n.a.
Zalora	1,100,000	3	5
GoShop	425,600	6	3
eBay	287,900	7	7
Decathlon	262,100	n.a.	n.a.
Qoo10	259,100	17	9
Applecrumby	239,700	n.a.	n.a.
Sephora	225,200	5	6

Source: IPrice (2021) <https://iprice.my/insights/mapofecommerce/en/>

Reportedly, the top countries for online shopping are China, Singapore, Japan, the USA and South Korea (Austrade, 2020). Shopee outstrips the others in the market according to web visits. Data on alternative metrics such as the number of merchants or revenue are unavailable although these can provide a more accurate understanding of the market share of each of these players and the market structure of e-commerce. Shopee also benefitted over the COVID period because the government gave RM150 & RM500 Shopping Vouchers at Shopee for youths aged 18–20 years and students at private and public institutions of higher learning under the e-belia initiative.[3]

Drivers of E-commerce Growth

The literature on the enablers or factors that foster e-commerce growth and/ or adoption focuses primarily at firm-level data, rather than the country-level information. Firm-level analysis centre on SMEs using the Technology-Organization-Environment (TOE) framework.[4] Country-level analysis, including multiple-country studies, is less common. Ho et al. (2007) combined endogenous and exogenous growth model to assess the key drivers of e-commerce growth for 17 European countries over a five-year period from 2000 to 2004. The ten key variables used in their model are derived from a literature review focussing mainly on empirical studies. These are GDP, geography, demography, urbanisation, info infrastructure, cost of on-line shopping, adequate resources, cosmopolitanism, education, and human capital. Their panel regression analysis indicates B2C e-commerce growth within a country is driven by internal drivers, with external factors indicating the possibility of regional contagion effects.

Subsequent studies at the macro-level also used the literature to derive the key e-commerce or digital growth enablers. Pasadilla et al. (2017) also summarised from the empirical literature the three main enablers which is namely infrastructure, factors facilitating supply and demand. Each of these have sub-indicators. Thus, infrastructure had ten sub-indicators representing access, speed, cost, and architecture. Supply variables were labour, capital and economic and regulatory environment, while demand variables were digital literacy, availability and use of digital payments, as well as indicators that capture the network externality benefits. These variables were then used to construct an index of digital enablers for ranking APEC countries. In that index, Singapore was ranked 3rd, while Malaysia was 13th, behind Singapore and ahead of the rest of ASEAN member states (AMS) who are members of Asia-Pacific (APEC) as well.

Kinda (2019) highlights internet access, secure internet servers to safeguard payments and personal information, availability of payment methods, reliable delivery services for physical goods and legal and institutional infrastructure as the key enabling factors for facilitating e-commerce growth. For Latvia, Gudele and Jekabsone (2020) identified the key enablers to be economic, technological, social, regulatory, psychological and political factors, although the sub-factors inside each of these enablers are not discussed nor measured.

Apart from these studies, there are two indexes that pertain to the development of e-commerce. They are UNCTAD's B2C E-Commerce Index[5] which provides a convenient (and partial) measure of factors that drive B2C e-commerce and the European Centre for International Political Economy (ECIPE)'s Digital Trade Restrictiveness Index (DTRI). The former consists of three indicators that are highly related to online shopping and for which there is wide country coverage such as account ownership at a financial institution or with a mobile-money-service provider (percent of population ages 15+); individuals using the internet (percent of population); Postal Reliability Index; and secure internet servers (per 1 million people). The latter incorporates four major classes of restrictions: (i) fiscal restrictions and market access, (ii) establishment restrictions, (iii) data restrictions, and (iv) trading restriction.[6]

Based on the literature review, we adapt Ho et al.'s model as it applied growth models for analysing e-commerce growth. The suggested variables in Ho et al.'s model are grouped into three main categories, that is economy, environment, and technology for ascertaining the drivers to e-commerce growth (see Figure 9.5). For the economy, we use variables that indicate the endowment of a country for conducting e-commerce while the environment indicates the operating environment in the country for e-commerce. Technology shows the info infrastructure available for conducting and facilitating e-commerce. There is, however, insufficient data for conducting a regression analysis on Malaysia alone. Neither is there enough comparable data on e-commerce to run a panel analysis on selected countries in ASEAN. The drivers of e-commerce-growth in Malaysia are therefore benchmarked against the best in the region.

Economy

The GDP per capita, geography, demography, urbanisation and digital talent for supporting e-commerce growth in a country are examined in this section. In Figure 9.6, the GDP per capita of Malaysia is second among the ASEAN-6 for the period shown, indicating that Malaysia has a larger potential for e-commerce growth since a larger market size will provide market opportunities and profit potential for e-commerce. For example, the most on-line fast-moving consumer goods (FMCG) sales are generated from nine of the world's ten largest economies (The Nielson Company 2018).

Malaysia is the fourth smallest country in terms of land area, behind Singapore, the Philippines and Vietnam. Geographically, archipelagic states[7] like the Philippines and Indonesia pose severe logistic challenges for e-commerce, especially in last-mile logistics. Malaysia's geography is less challenging for logistics as it only has two significant regions, namely Peninsular Malaysia and East Malaysia.

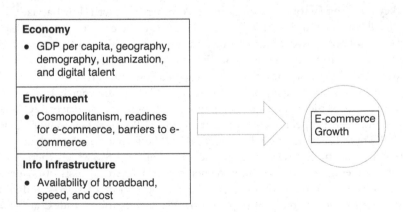

Figure 9.5 Key Enablers of E-commerce Growth in Malaysia.
Source: Adapted from Ho et al. (2017).

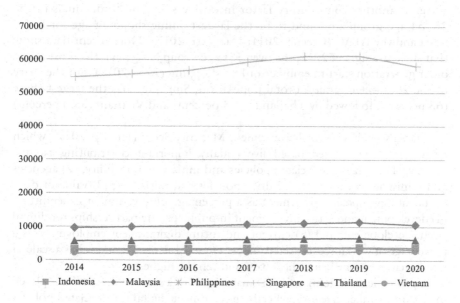

Figure 9.6 GDP Per Capita (Constant 2015 US$), 2014–2020

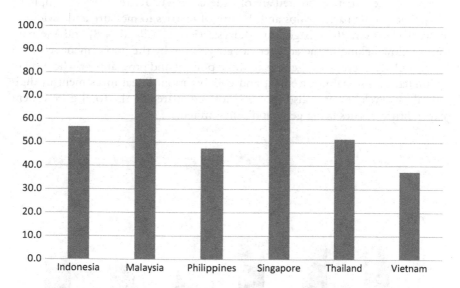

Figure 9.7 Urban population share in total population, 2020 (percent).

In general, a higher population density will lower logistics and infrastructure costs, thereby offering greater advantages for e-commerce. Thus, for example, Cheng et al. (2021) found that higher population density leads to a higher number of parcel deliveries. Malaysia's urban population as a percentage total population is the second largest among the ASEAN-6, in 2020 (Figure 9.7).

Age is another contributory factor in online shopping. Individuals in the 25–64 age group are found to be the largest online shopping age for Indonesia and the APAC (Greene, 2014; Hasyyati, 2017). More recent data show that since the start of 2015, the over-55s age group has grown the most as the older generation start to embrace online shopping (Read 2019). For the share of the 25–64 age group in total population, Singapore has the largest share (63 percent), followed by Thailand (57.3 percent) and Vietnam (55.3 percent) (Figure 9.8).

Huawei's Digital Talent Development Maturity Score (Huawei 2021), which is an index constructed based five equally important contributing factors, namely, (1) digital talent related policies and initiatives availability; (2) agencies and ministries collaboration to drive policies and initiatives; (3) value of digital talent development investment as a percentage of government expenditure; (4) development and active involvement in public-private partnerships for digital talent development; and (5) current and future digital talent employment as a percentage of workforce, is used to assess Digital Talent. The index has a scale of one to ten, with one being the lowest and ten the highest.

Singapore leads in terms of DTD maturity with a score of 8.4, and it is placed in the frontrunner category, whereby government digital-talent-related policies are driven and collaborated through a few ministries within its government agencies while the initiatives created are of wide coverage. There is also an emphasis on public-private partnerships and the use of metrics to measure and review its efficacy (Huawei, 2021, page 56). Malaysia (5.0), Thailand (4.8) and Vietnam (3.6) follow behind in the adopter category, whereby the government relies on a handful of government agencies to drive policies and programs or where there is limited ministerial involvement and collaboration. Fiscal investment in digital talent development is also relatively low compared to the total government expenditure besides being geographically limited.

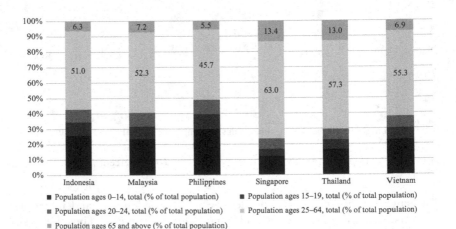

Figure 9.8 Population by Age Group.

Indonesia and the Philippines are placed in the last category (starters), with a respective score of 4.2 and 2.0. In this category, the government relies on international private enterprises to develop digital talent in its country, and government involvement is limited. There is insufficient digital talent fiscal investment per capita while policies and initiatives are narrow, as they do not take into account varied career phases, ages, skill-sets and geography. Overall, in terms of the economy, Singapore leads for the variables discussed, followed by Malaysia, except in the case of geographical and demographical factors.

Environment

Cosmopolitanism matters in e-commerce development as it indicates the willingness of an individual to go beyond the boundaries of one's community to engage in different cultural experiences (Han and Won, 2018). Thus, consumers with cosmopolitan values will be more willing to open up to cross-border consumption, and experience global brands and products, as available in cross-border e-commerce. Kandogan (2019) used actual macro cross-border trade data rather than individual purchase intentions to measure the degree of consumer ethnocentrism versus cosmopolitanism in each country for the G20 and emerging markets. He found that Malaysia is open to cosmopolitanism, but Indonesia, the Philippines and Thailand veer towards ethno-centrism. It should be noted that Singapore and Vietnam are not included in the sample.

The readiness to embrace or restrict e-commerce growth is represented by UNCTAD's B2C index and the ECIPE's DTRI represents the openness to trade in e-commerce. Another restriction on e-commerce activities is the rise of on-line frauds, with e-commerce being the most reported scam in Singapore in 2020 (Bose, 2021). Interpol (2021) reports a rise of cybercrimes in ASEAN, including e-commerce data interceptions and cyberscams. ITU's Global Cybersecurity Index (GCI) is used to assess the national cybersecurity capacity of a country. Each country's level of development or engagement is assessed along five pillars – (i) Legal Measures, (ii) Technical Measures, (iii) Organisational Measures, (iv) Capacity Development and (v) Cooperation. The five are then aggregated into an overall score.

UNCTAD's B2C e-commerce index, which measures an economy's preparedness to support online shopping, is shown in Figure 9.9. All countries have improved in their index value over time. For both years, Malaysia is ranked behind Singapore but ahead of the other ASEAN member states for this index indicating that Malaysia is more prepared than her ASEAN neighbours, with the exception of Singapore.

A closer look at Malaysia's changes in scores for the three indicators used to construct the B2C indicator shows that the biggest improvement is found in the share of individuals with an account, while the smallest change is secure internet servers per million population, which also has the lowest score out of the three indicators in 2020 (Figure 9.10).

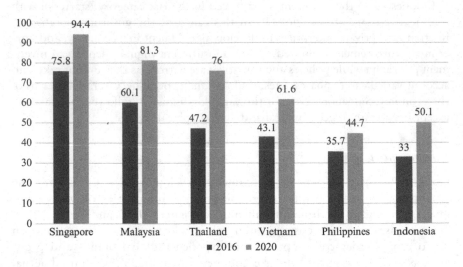

Figure 9.9 UNCTAD B2C E-commerce Index Value.

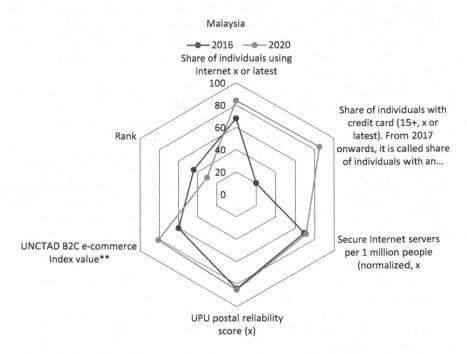

Figure 9.10 Malaysia Score Breakdown in UNCTAD B2C E-commerce Index.

Malaysia's environment may restrict DT. For this, the DTRI is used, as it measures how countries in the world restrict DT. The overall DTRI ranges from 0 (i.e. completely open) to 1 (i.e. virtually restricted) with increasing values representing higher levels of DT costs for businesses. The DTR Index shows that Malaysia

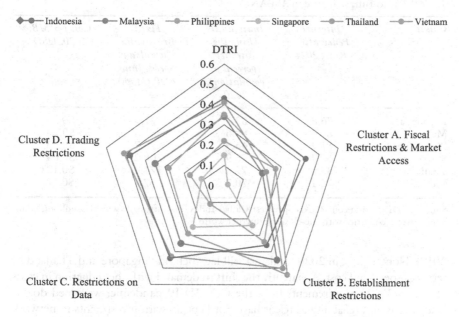

Figure 9.11 Digital Trade Restrictiveness Index (DTRI) of SEA Countries.

is more restrictive compared to Singapore and the Philippines (Figure 9.11). Malaysia is less restrictive than the Philippines for one dimension, namely fiscal restrictions and market access. Of the four dimensions, Malaysia is most restrictive in the establishment dimensions.

The 2020 ITU Global Cybersecurity Index ranks Singapore number one in terms of capacity to manage cybersecurity, in the APAC, followed by Malaysia (2), Indonesia (6), Vietnam (7), Thailand (9) and the Philippines (13) (ITU 2021). Overall, Singapore's environment leads in the region, for facilitating e-commerce growth, followed by Malaysia, except in restrictiveness where it is behind Singapore and the Philippines and Thailand for some dimensions.

Info Infrastructure

Based on Table 9.3, Malaysia, among the ASEAN-6, has the best info infrastructure for supporting e-commerce in terms of internet penetration and individuals using the internet. Nevertheless, fixed broadband download speed is faster in Singapore and Thailand. Likewise, cost per MB is higher in Malaysia relative to Singapore, Thailand and Vietnam. According to World Bank (2018), the high cost of fixed broadband internet services is partly driven by limited competition, with Telekom Malaysia Bhd (TM) having a significantly larger market share than the leading firms in other countries. In 2018, Malaysia implemented the Mandatory Standard on Access Pricing (MSAP), which regulated the prices and terms for alternative Internet Service Providers (ISPs) to access the wholesale broadband capacity. This led to a subsequent reduction in broadband prices (Siddhartha and Record,

Table 9.3 Info Infrastructure in ASEAN-6

Countries	Internet Penetration Rate, 2021	Individuals Using the Internet (percent of population), 2020	Fixed Broadband Download Speed, June 2020 (Mbps)	Cost per MB, 2020, USD
Indonesia	76.8	53.7	21.3	$0.56
Malaysia	89.0	89.6	81.5	$0.24
Philippines	81.9	46.9	23.7	$0.75
Singapore	87.7	75.9	208.2	$0.04
Thailand	83.6	77.8	171.4	$0.12
Vietnam	77.4	70.3	57.5	$0.17

Source: WDI, Statista.com, Ookla Speedtest Global Index, and https://www.atlasandboots.com/remote-work/countries-with-the-cheapest-internet-world/

2019). Nonetheless, in 2020, Malaysia still lags behind Singapore and Thailand in terms of speed and cost. Apparently the shift in demand with the reduction in costs as well as WFH requirements from the COVID-19 pandemic, worsened download speeds and reliability as telcos have not kept up with investments in network capacity (Edge Malaysia 2021). Hence, the quality of info infrastructure in terms of speed and cost is still less attractive compared with some other AMS.

Challenges and Policy Implications

In the DTRI, Malaysia had the highest score (more restrictive) for establishment restrictions, which can be in the form of equity requirements or the market structure of various services sub-sector that can affect the growth of the e-commerce market. At the same time, the need for market access has pressed for the inclusion of increasing digital commitments in FTAs. All three factors can affect the future development of e-commerce in the country.

Equity Constraints

The e-commerce ecosystem covers mainly services such as telecommunication services, marketing, platform players, logistics (including all modes of transportation), warehousing, distribution and e-fulfilment centres as well as real estate. Based on OECD's FDI regulatory restrictiveness index (Table 9.4), Malaysia's index is higher (more restrictive) than the OECD average, Singapore and Vietnam. More importantly, the scores are all higher in the relevant services subs-sectors for e-commerce compared to the overall score which includes manufacturing, which is relatively more open than the services sector.

Although the Companies Act, 2016 does not stipulate any equity conditions on Malaysian incorporated companies, the government encourages joint-ventures/partnership between Malaysian and foreign investors.

In services, specific equity conditions are imposed for specific approvals, operating licences, permits or registrations by the regulating Ministries/Agencies, depending on the activities undertaken. Therefore, the equity constraints in terms of the cap for foreign ownership and Bumiputera ownership requirements vary according to sub-sectors in services. For example, in the logistics sector, there is no equity condition for ordinary warehouses, but public bonded warehouses must have at least 30 percent Bumiputera equity while private bonded warehouses have no equity conditions. Freight forwarders with an Integrated Logistics Services (IILS) status from MIDA, allows 100 percent foreign equity ownership (MIDA undated).

Bumiputera equity requirements are revised periodically, which lend considerable uncertainty for investors with licenses that need annual renewal. To illustrate, in January 2021, the Ministry of Finance (MoF) issued a letter stating that all customs related licences, such as freight-forwarders/customs agents will be required to comply with reportedly a 51 percent Bumiputera equity requirements by end 2021 (Hsian & Co, 2021). Although the deadline for compliance to the new requirements have been extended to 31 December 2022, in response to queries and objections from the stakeholders and the public at large, this type of changes heightens uncertainty in the investment regime.

Uncertainty also reigns when equity requirements are not stipulated but given on a case-by-case basis. In the telecommunication sector, licenses are needed to operate, with four categories of licenses that are market specific. These can be individual or class licenses. WTO (2017) noted that although there are no foreign equity restrictions on class licences, individual licences are assessed on a case-by-case basis. This implies that equity conditions can be imposed as licence conditions. In general, 49 percent foreign equity is allowed for a network facility or service provider, together with another condition, 30 percent Bumiputera equity requirement. These equity restrictions apply for private listed companies and not public listed companies.

Table 9.4 OECD FDI Regulatory Restrictiveness Index, 2020

Sectors	OECD Average	Indonesia	Malaysia	Philippines	Singapore	Thailand	Viet Nam
Distribution	0.017	0.555	0.470	0.153	0.013	0.073	0.126
Wholesale	0.017	0.355	0.370	0.065	0.013	0.073	0.045
Retail	0.018	0.755	0.570	0.240	0.013	0.073	0.208
Transport	0.204	0.515	0.296	0.655	0.129	0.378	0.522
Communications	0.079	0.380	0.375	0.650	0.063	0.490	0.583
Fixed telecoms	0.085	0.380	0.375	0.635	0.063	0.540	0.583
Mobile telecoms	0.073	0.380	0.375	0.665	0.063	0.440	0.583
Financial services	0.032	0.222	0.319	0.115	0.021	0.456	0.118
Real estate investment	0.158	0.105	0.300	0.525	0.167	0.373	0.237
Total FDI Index	0.063	0.347	0.257	0.374	0.059	0.268	0.130

Source: OECD.

In the interest of transparency and certainty, all equity conditions need to be clearly stated, with no sudden changes that can derail and an investor's plans, regardless of domestic or foreign.

Competition

The telecommunication sector still has substantial government interest, including shareholdings by government-linked investment companies (GLICs). The privatised national telecommunication company TM is still majority-owned by GLICs, including Khazanah National Bhd, Employees Provident Board, AmanahRaya Trustees Bhd.[8] TM dominates the fixed-line market as the incumbent operator before privatisation, with as estimated market share of 95 percent. Although five additional licences were awarded to operate the fixed-line market, the prohibitive cost of building a fixed line network from scratch effectively deterred new entrants (Venugopal, 2003), thereby making TM a natural monopoly in this segment. The Malaysian Communications and Multimedia Commission (MCMC) as the regulator based on the powers provided for in the Malaysian Communications and Multimedia Commission Act 1998, which includes provisions for the regulation of market power in this sector, has to exercise economic regulation, by ensuring that TM does not charge monopolistic pricing in providing access to fixed line and last mile.

Since mobile broadband market is less capital-intensive, more competition was injected with eight providers awarded the 4G spectrum. Of these, in the 2020 subscription market share, Maxis commanded the highest share of 28.2 percent, followed by Digi (22.8percent) and Celcom (18.3 percent) (MCMC, 2020). The remainder was from U Mobile and others/Mobile Virtual Network Operators (MVNOs), with a 16.4 percent and 14.2 percent share, respectively. Axiata and Celcom (wholly owned subsidiary of Axiata) have substantial shares held by GLICs, notably, Khazanah National Bhd.

Thus, privatisation and liberalisation in the Malaysian telecommunication market have led to only a partial improvement in competition in certain segments of the market. This is unlike Thailand where the legacy duopolists have become small market players and newer private companies have entered and gained bigger market shares (Pornchai and Nasarudin, 2020). In November 2021, Celcom Axiata and Digi made a formal application for a proposed merger, thereby reducing the number of players in the market. The telecommunication market can be considered as oligopoly market where tacit collusion is possible, making a collective dominant possible (Nasarudin et al., 2014). Given that there is also common ownership with the GLICs having shares in these companies, MCMC's role as regulator has to enhance transparency in its decision-making through public consultations and making known the outcomes of any investigations on competition issues in this sector to ensure that there is no abuse of any possible dominant position that may emerge due to common ownership.

The on-going reported impasse in December 2021 between the mobile operators and Digital Nasional Berhad (DNB), the state agency tasked with deploying 5G, illustrate these operators working together purportedly to press for transparency and clarity on pricing issues. But MCMC, the regulator, which also regulates the market power in this sector, had already earlier in February 2021, issued a clarification that DNB will be offering services via wholesale that is transparent and regulated, while service providers will have open, fair and equal access to the new network. This new model via wholesale is to ensure a shift from infrastructure-based competition to services-based competition (MCMC 2021). MCMC therefore as regulator has to step in to resolve the impasse by ensuring what has been stated will be fulfilled. Pricing also has to take into account affordability for the consumer and the SMEs that are supposed to benefit from the enhanced speed for upgrading their businesses so that Malaysia can catch up with some of her regional peers in this (Table 9.3).

There is also increasing concern that anti-competitive behaviour may emerge in the online space (OECD, 2018). In Malaysia, numerous problems encountered by consumers, including unfair trade practices (Shivani Sothirachagan, 2021). The Federation of Malaysian Consumers Association (FOMCA) and Consumers' Association of Penang (CAP) have also received complaints on the issue of competition (FMT 2021). There is a need to ascertain if any of the shopping platforms have a dominant position and if there is any abuse of a dominant position since Section 10 of the Competition Act prohibits abuse of a dominant position. Data are needed to measure dominant position, but such data are not available for public measurement nor scrutiny as platform providers are the holders of these data. The Malaysian Competition Commission has to move beyond the brick and mortar world to the virtual world for addressing the new issues emerging from an increase in e-commerce activities in the country.

Digital Commitments and the Need for Better Data

Despite the scarcity of digital data, the growing importance of such data, especially cross-border data, has escalated the inclusion of digital commitments in Free Trade Agreements as countries press for greater market access, trade facilitation in the digital world and rules and regulations to govern cross-border DT. Malaysia is a party to the WTO Joint Statement Initiative (JSI) on E-commerce that supports the continuation of the multilateral e-commerce moratorium in fostering certainty and predictability for businesses. The December 2021 statement from the WTO announced a good convergence on eight articles, namely online consumer protection; electronic signatures and authentication; unsolicited commercial electronic messages; open government data; electronic contracts; transparency; paperless trading; and open internet access (WTO, 2021). The group is working towards the successful conclusion of the JSI in the near future.

Malaysia is also party to the ASEAN-Australia-New Zealand FTA, the ASEAN E-commerce agreement, Regional Comprehensive Economic Partnership Agreement (RCEP) and the Comprehensive and Progressive Agreement for Trans-Pacific Partnership (CPTPP). These agreements have a chapter on e-commerce and the provisions in each agreement have increased over time for the parties to each of these agreements (Tham, 2021). However, the CPTPP has more provisions on e-commerce, despite being an older agreement than the RCEP. The RCEP also has a provision on dialogue on e-commerce which includes anti-competitive practices, and on-line dispute resolution (Article 12.16). As noted by Wu et al. (2021), although there is no specific provision on enhanced competition enforcement, nevertheless competition rules are relevant for specific provisions such as protecting online consumers and the creation of a conducive environment for e-commerce.

There is increasing pressure to make more and deeper digital commitments as ASEAN moves towards negotiating ASEAN Digital Economy Framework in the near future. These deeper provisions include, for example, electronic invoicing, electronic payments, cooperation on competition policy, submarine telecommunications cable, location of computing facilities for financial services, data innovation, open government data, digital identities, standards and conformity assessment for DT, artificial intelligence and fintech cooperation, all of which require more and better data to understand the magnitude and importance such issues as well as to ascertain the impact of binding commitments in FTAs on the future development of e-commerce. Although Malaysia is compiling data on e-commerce and the digital economy, there are still large gaps, especially firm-level data which is needed for facilitating better research and providing evidence-based policy suggestions.

Conclusion

Malaysia's e-commerce has grown over the years, with the COVID-19 pandemic accelerating this shift. The shift is facilitated by favourable economic factors such as relatively high per capita income, young urban population, and a digital talent maturity index that is just behind Singapore. Malaysia also faces a relatively favourable environment for fostering e-commerce growth according to UNCTAD's B2C index and the 2020 ITU Global Cybersecurity Index. However, Malaysia is relatively less well placed compared with some of her ASEAN neighbours, in terms of some of the DT restrictions and broadband speed and cost.

In moving forward, Malaysia's relatively higher restrictions on establishments is also reflected in the FDI restrictiveness index. The restrictions in establishments are in the form of equity constraints such as cap on foreign equity and Bumiputera equity requirements. These requirements are not homogeneous across the different service sub-sectors that cover e-commerce. In particular, Bumiputera equity constraints are periodically revised, lending uncertainty for domestic and foreign investors. In the telecommunication sector, even without any equity constraints, equity restrictions can be imposed as conditions for licenses given out.

In both the telecommunication sector and e-commerce platforms, there are possibilities of dominant positions, or collective dominant positions due to the nature of the market structure where there are only a small number of players or a few large players as well as a large number of much smaller players. Regulators have an important role to play in terms of ensuring consumers are protected in the digital world nor abuse of any potential collective dominant position. Finally, as countries move on to used FTAs to govern the DT, more and better data is needed to understand the magnitude of the issues and the impact of these binding commitments on the future development of DT and the future digital economy.

Notes

1 The authors would like to thank Dr. Wan Khatina Nawawi, former economic advisor to the Malaysian Competition Commission, for her inputs on competition policy in the telecommunication sector. The usual caveat applies
2 E-commerce transaction is defined to be the sale or purchase of goods or services, conducted over computer networks by methods specifically designed for the purpose of receiving or placing of orders. Method of payment and the ultimate delivery of the e-commerce goods or services might be done through computer network/internet or traditionally
3 See https://afterschool.my/articles/government-through-e-belia-is-giving-rm150-rm500-shopping-vouchers-at-shopee-for-youth-aged-18–20-years-students-at-ipta-ipts-nationwide <Accessed 20 December 2021>
4 See literature review in Kam and Tham (2022).
5 See UNCTAD Technical Notes on ICT for Development, https://unctad.org/system/files/official-document/tn_unctad_ict4d14_en.pdf
6 See https://ecipe.org/wp-content/uploads/2018/05/DTRI_FINAL.pdf
7 This refers to island countries that consist of an archipelago. There are 22 island countries that have sought claim to archipelagic status based on the 1982 UN Convention on the Law of the Sea provisions.
8 See https://tm.listedcompany.com/capital_structure.html <Accessed 16 December 2021>

References

Austrade (2020). *E-commerce in Malaysia: A Guide for Australian Business.* Canberra: Commonwealth of Australia.

Bose, S. (2021). "COVID-19: The Evolution of Scams in Asia-Pacific". *Singapore Business Review.* https://sbr.com.sg/information-technology/commentary/covid-19-evolution-scams-in-asia-pacific <Accessed 14 December 2021>

Cambridge Econometrics (2020). *Understanding and Measuring Cross-Border Digital Trade Final Research Report.* https://assets.publishing.service.gov.uk/government/uploads/system/uploads/attachment_data/file/885174/Understanding-and-measuring-cross-border-digital-trade.pdf <Accessed 14 December 2021>

Cheng, C., Sakai, T., Alho, A., Cheah, L. & Ben-Akiva, M. (2021). "Exploring the Relationship between Locational and Household Characteristics and E-Commerce Home Delivery Demand". *Logistics,* 5, 29. https://doi.org/10.3390/ logistics5020029

Digital News Asia (2021). *Mdec Sets Lofty Target for E-Commerce Adoption By 2025.* https://www.digitalnewsasia.com/business/mdec-sets-lofty-target-e-commerce-adoption-2025 <Accessed 14 December 2021>

Edge (2021). *Cover Story: With Single Shared Infrastructure, 5G Costs to Malaysians Should be Half Current 4G Price (from RM2.GB to RM1/GB).with Telcos Making same Absolute Profits and More.* December 11 2021. https://www.theedgemarkets.com/article/cover-story-single-shared-infrastructure-5g-costs-malaysians-should-be-half-current-4g-price <Accessed 14 December 2021>

European Centre for International Political Economy (ECIPE) (2018). *Digital Trade Restrictiveness Index*. https://ecipe.org/wp-content/uploads/2018/05/DTRI_FINAL.pdf <Accessed 9 December 2021>

FMT (2021). *Consumer Groups Call on MyCC to Probe E-commerce Giant*. https://www.freemalaysiatoday.com/category/nation/2021/05/06/consumer-groups-call-on-mycc-to-probe-e-commerce-giant <Accessed 20 December 2021>

Gloria, O., Pasadilla, A. W., Chiang, T. W. & Adrian W. (2017). "Facilitating Digital Trade for Inclusive Growth Key Issues in Promoting Digital Trade in APEC". *Issue Paper No. 12*. Singapore: APEC Policy Support Unit.

Gomez, E.T., Tham, S.Y., Li, R. & Cheong, K.C. (2020). *China in Malaysia: State-Business Relations and the New Order of Investment Flows*. Singapore: Springer Nature Singapore Pte. Ltd.

Google, Temasek, Bain & Company (2021). *E-Conomy SEA 2021. Roaring 20s: The SEA Digital Decade*. https://www.bain.com/globalassets/noindex/2021/e_conomy_sea_2021_report.pdf <Accessed 16 December 2021>

Greene, Shawn (2014). *E-Commerce Trends and Development in Asia-Pacific*. https://www.asiabriefing.com/news/2014/06/e-commerce-trends-developments-asia-pacific/ <Accessed 9 December 2021>

Gudele, I. & Jekabsone, I. (2020). Factors Contributing to the Development of E-Commerce by the Degree of use in Latvia. *European Integration Studies* No. 14 / 2020, pp. 207–216. https://doi.org/10.5755/j01.eis.1.14.26385

Han, C. M. & Won, S. B. (2018). "Cross-country Differences in Consumer Cosmopolitanism and Ethnocentrism: A Multilevel Analysis with 21 Countries". *Journal of Consumer Behaviour*, *17*(1), e52–e66. https://doi.org/10.1002/cb.1675

Hasnah, M. & Mazreha Y. (2019). "E-commerce: The Contribution to the Malaysia's Economy". *Paper presented at Asia–Pacific Economic Statistics Week 2019*. https://www.dosm.gov.my/v1/uploads/files/7_Publication/Technical_Paper/Paper_APES/2019/2.%20E-Commerce%20The%20Contribution%20To%20The%20Malaysia's%20Economy.pdf<Accessed 15 December 2021>

Hasyyati, A. N. (2017). "Demographic and Socioeconomic Characteristics of E-commerce Users in Indonesia". *ADBI Working Paper, No. 776*. Tokyo: Asian Development Bank Institute (ADBI).

Herbert S. F. (2018). *Digital Trade Definition*. https://globalaccesspartners.org/HSF-Digital-trade-definition.pdf <Accessed 14 December 2021>

Hew, L. (2021). "MDEC Targets 875,000 MSMEs to Adopt e-commerce by 2025". *Edge Markets*, April 27, from https://www.theedgemarkets.com/article/mdec-targets-875000-msmes-adopt-ecommerce-2025 accessed on April 20.

Ho, S.C., Kauffman, R. J., & Liang, T. P. (2007). "A Growth Theory Perspective On B2C E-commerce Growth in Europe: An Exploratory Study". *Electronic Commerce Research and Applications*, 6(2007), 237–259.

Hsian & Co (2021). *Bumiputera Equity Requirement for Freight Forwarders*. https://www.lexology.com/library/detail.aspx?g=bd17f816-485a-44f5-8271-43f3d3182c64 <Accessed 13 December 2021>

Huawei (2021) (2022). "Bridging the Gap: Matching Digital Skills and Employability Pipeline". *Digital Talent Insight*. https://www-file.huawei.com/-/media/corp2020/pdf/event/1/2022_digital_talent_insight_asia%20pacific.pdf?la=en <Accessed 13 December 2021>

IPrice (2021). *Insights: The Map of E-commerce in Malaysia*. https://iprice.my/insights/mapofecommerce/en/ <Accessed 13 December 2021>

Interpol (2021). *Asean Cyberthreat Assessment 2021: Key Cyberthreat Trends Outlook from the Asean Cybercrime Operations Desk.* https://www.interpol.int/content/download/16106/file/ASEAN%20Cyberthreat%20Assessment%202021%20-%20final.pdf <Accessed 13 December 2021>

International Telecommunication Union (ITU) (2021). *Global Cybersecurity Index 2020.* https://www.itu.int/dms_pub/itu-d/opb/str/D-STR-GCI.01-2021-PDF-E.pdf <Accessed 14 December 2021>

Kam, A. J. Y. & Tham, S. Y. (2022). "Barriers to E-commerce Adoption in Developing Country: Evidence from the Retail and Food and Beverage Sector in Malaysia". *Asian-Pacific Economic Literature,* 36(2), 32–51. https://doi.org/10.1111/apel.12365

Kandogan, Y. (2019). "Using Macro Cross-Border Trade Data to Better Understand Micro-Level Country of Origin Effects", *Thunderbird International Business Review,* 62(2), 213–226.

Kinda, T. (2019). "E-commerce as a Potential New Engine for Growth in Asia", *IMF Working Paper,* WP/19/135.

Malaysia (1996). *Seventh Malaysia Plan 1996–2000.* Putrajaya: Government Printers.

Mazreha, Y. & Norazlin, M. (2021). "Malaysia E-commerce: Performance and Expectation". *Presentation at Asia-Pacific Stats Café Series 2021.* https://www.unescap.org/sites/default/d8files/event-documents/Malaysia_e-Commerce_Performance_Expectation_Stats_Cafe_23Aug2021.pdf <Accessed 14 December 2021>

MCMC (2020). *Industry Performance Report 2020.* https://www.mcmc.gov.my/skmmgovmy/media/General/pdf/MCMC-IPR_2020.pdf <Accessed 14 December 2021>

MCMC (2021). *Further Clarification on Approach to 5G Deployment in Malaysia.* https://www.mcmc.gov.my/en/media/press-releases/syarikat-tujuan-khas-untuk-memacu-pelaksanaan-5g-m <Accessed 14 December 2021>

MIDA undated. *Malaysia: Investment in Services - Logistic Services.* https://www.mida.gov.my/wp-content/uploads/2020/07/20191220163458_BOOKLET-4-LOGISTICS-SERVICES.pdf <Accessed 14 December 2021>

Nasarudin A. R., Haniff, A. & Juriah A. J. (2014). "Regulating Competition in the Malaysian Telecommunication Sector: A Need for a New Approach?". *International Journal of Public Law and Policy,* 4(4), 403–419.

OECD undated. *Digital Trade.* https://www.oecd.org/trade/topics/digital-trade/ <Accessed 14 December 2021>

OECD (2011). *OECD Guide to Measuring the Information Society 2011.* Paris: OECD Publishing. http://dx.doi.org/10.1787/9789264113541-en

OECD (2018). *Implications of E-commerce for Competition Policy - Background Note.* https://one.oecd.org/document/DAF/COMP(2018)3/en/pdf <Accessed 14 December 2021>

Pornchai, W. & Nasarudin, A. R. (2020). "Regulatory Frameworks for Reforms of State-owned Enterprises in Thailand and Malaysia". *ADBI Working Paper Series.* No. 1122 April 2020. Tokyo: ADBI.

Read, A. (2019). *Ecommerce and the Demographics of Online Shoppers.* https://www.further.co.uk/blog/ecommerce-and-the-demographics-of-online-shoppers/<Accessed 9 December 2021>

Shivani S. (2021). Protecting Malaysian E-commerce Consumers. *Presentation at the Eleventh Meeting of the UNCTAD Research Partnership Platform 17–18 December 2020.* https://unctad.org/system/files/non-official-document/ccpb_RPP_2020_02_Present_Sivani_Sothrachagan.pdf <Accessed 20 December 2021>

Siddhartha, R. & Record, R. (2019). *Malaysia's Need for Speed: How Regulatory Action is Unleashing Ultrafast Internet.* https://blogs.worldbank.org/eastasiapacific/malaysias-need-speed-how-regulatory-action-unleashing-ultrafast-internet <Accessed 14 December 2021>

Tham, S. Y. (2020). Development of E-commerce in Malaysia. Chapter 7 in Lee, Cassey and Lee, Eileen (eds.), *E-commerce, Competition and ASEAN Economic Integration.* Singapore: ISEAS.

Tham, S. Y. (2021). *Digital Commitments in ASEAN's Free Trade Agreements.* Perspective *2021/163.* Singapore: ISEAS. https://www.iseas.edu.sg/articles-commentaries/iseas-perspective/2021-163-digital-commitments-in-aseans-free-trade-agreements-by-tham-siew-yean/ <Accessed 19 December 2021>

The Nielsen Company (2018). *Future Opportunities in FMCG E-commerce: Market Drivers and Five-Year Forecast.* https://www.nielsen.com/wp-content/uploads/sites/3/2019/04/fmcg-eCommerce-report.pdf <Accessed 9 December 2021>

Venugopal, K. (2003). *Telecommunication Sector Negotiations at the WTO: Case studies of India, Sri Lanka and Malaysia.* Paper presented at ITU/ESCAP/WTO Regional Seminar on Telecommunications and Trade Issues 28–30 October 2003, Bangkok, Thailand. https://citeseerx.ist.psu.edu/viewdoc/download?doi=10.1.1.111.2439&rep=rep1&type=pdf <Accessed 14 December 2021>

World Bank (2018). *Malaysia Economic Monitor: Navigating Change.* Washington DC: The World Bank Group. https://openknowledge.worldbank.org/handle/10986/29926 <Accessed 14 December 2021>

World Trade Organisation (WTO) (2017). *Trade Policy Review: Malaysia.* https://www.wto.org/english/tratop_e/tpr_e/s366_e.pdf <Accessed 14 December 2021>

World Trade Organisation (WTO) 2021. *WTO Joint Statement Initiative on E-commerce Statement by Ministers of Australia, Japan and Singapore.* https://www.wto.org/english/news_e/news21_e/ji_ecom_minister_statement_e.pdf <Accessed 14 December 2021>

Wu, P., Weng, C.X. and Joseph, S.A. (2021). "Crossing the Rubicon? The Implications of RCEP on Anti-monopoly Enforcement on Dominant E-commerce Platforms in China", *Computer Law & Security Review, 42,* 1–15. https://doi.org/10.1016/j.clsr.2021.105608.

10 Digitalization and the Gig Economy

Ecosystem System Supporting Consumer Services

Khairul Hanim Pazim, Yosuke Uchiyama, Fumitaka Furuoka, Beatrice Lim, and Jingyi Li

Introduction

The rise of the digital economy has made it easier for consumers to organize their lives. For instance, commuting may be a lot easier with a car booked through Grab. The global impact of the digital economy will be in the range of USD 3.1 trillion with a potential growth rate of 31 percent according to the study where Malaysia has the potential to see a contribution of up to USD 14 billion to its GDP by 2025.

Before the pandemic, the digital economy had already started expanding well before. In 2019, Malaysia has more than 80 local firms and is still growing. The number of users in Malaysia's ride-hailing and taxi segment is expected to reach 7.7 million by 2026. In terms of e-payment usage, Touch 'n Go was used for e-payment transactions by approximately 82.4 percent of e-payment users. In the same survey, 91 percent of respondents said they had used e-payment methods to complete a transaction. The spread of the COVID-19 pandemic has resulted in the Malaysian government issuing a Movement Control Order (MCO) in March 2020 to combat COVID-19, which has affected the SE.

Lockdown and quarantine in response to a public health emergency may have a significant impact on citizens' individual choices, daily habits, and behaviours. The pandemic's economic effects have caused business and labour market disruption. The firms' actual operations were not only halted, but digital services were also impacted. For example, digital services in hospitality and transportation have suffered as a result of travel restrictions. As a result of the pandemic, demand for popular SE services like Uber, Lyft, and other ride-hailing applications has dropped dramatically, resulting in a downward spiral in rides and revenues. Numerous employees have been laid off, SE firms' value has plummeted, and many service providers are left with no choice but to cease operations (Hossain, 2021).

Some major drivers of the digital economy, which are based on shared assets, such as Uber and Grab, are also main players in the gig economy which are characterized by flexible and temporary work. Interestingly, these hybrid firms which are engaged in both the digital economy and gig economy have enjoyed higher profits under the pandemic rather than during normal times (Figure 10.1). For

DOI: 10.4324/9781003367093-10

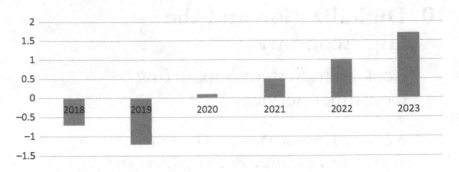

Figure 10.1 Profits of Grab from 2019 to 2023 (US$ Billion).

example, the dominant platform firm in Southeast Asia, Grab, made much better performance under the prolonged period of the lockdown and quarantine. The firm suffered from serious negative profits before the pandemic. However, its profits turned into positive in 2020 due to high profitability of delivery service. The platform firm expected to make handsome profits in the period of 2021–2023 (Kana 2021; Oi, 2021).

Although ride-hailing has declined during the pandemic, the importance of the digital economy has increased during the COVID-19 pandemic, because many sectors, such as healthcare and homecare, are closely associated with the SE, particularly in larger cities. COVID-19 pandemic has created some new opportunities for SE. Due to the fact that several countries are on lockdown, online shopping services have increased in popularity. Due to the closure of offices, the demand for freelance work has increased. As movie theatres have closed, video streaming services like Netflix, Disney, and Amazon Prime have risen to prominence as the primary source of entertainment. Thus, COVID-19 pandemic has been certainly changing the way consumers use digital online platforms, but it has also been increasing the importance of the digital economy in people's lives. The Malaysian government, under the Malaysian Digital Economy Blueprint, has identified the "GigUp" programme to improve the skills and employment status of gig workers in the digital economy by 2025 as part of the National Digital Strategy (Economic Planning Unit, Prime Minister's department, n.d.). At the same time, empirical research into the ecosystem and adaptability of consumers affected by these services should be further deepened, especially during the pandemic era.

The Shared Economy

The term "shared economy" is frequently used in academic literature referring to the field of sharing resources in an efficient way (Agarwal & Steinmetz, 2019). Although there is no consensus in the literature on the scope of shared economy (Barbu et al., 2018), it refers to an umbrella term for a variety of non-ownership forms of consumption activities. Although the terms "platform market," "gig economy," "collaborative economy" (Hamari et al., 2016), "sharing economy" (Cheng, 2016), "peer-to-peer economy" (Wirtz et al., 2019), "collective consumption" (Luri Minami et al., 2021), and "access-based consumption" (Bardhi & Eckhardt, 2012) may differ, they are commonly used interchangeably. Hamari et al. (2016) defined the shared economy as a new economic model that emphasizes the peer-to-peer-based activity of obtaining, giving, or sharing access to goods and services, coordinated through community-based online services to maximize the use of underutilized resources (Wirtz et al., 2019). Examples include Airbnb, which hosts the most homestay properties in the world without owning a single physical asset, and e-hailing firms like Grab and Uber, which provide transportation services without owning any vehicles (Table 10.1).

In general, the stakeholders involved in SE are SE firms or platforms, service providers, service receivers (customers), and regulatory bodies (Hossain, 2021). According to Hossain (2021), the SE differs from other businesses, because service providers, who own key resources like vehicles and properties, are not employees of the parent firm. Businesses that run such platforms typically act as

Table 10.1 Major Platform Firms and Their Activities in the Shared Economy

Platform Name	Industry and Main Competition	Geographic Regions of Operation	Supply Side Participants
Airbnb	Accommodation: Hotels, short-term rental firms	Global	Hosts, i.e., individuals with under-used property
Uber	Vehicle hire: Taxi firms, other ride-sharing platforms	Global	Drivers, i.e., individuals with cars
Grab • taxis (GrabTaxi) • private car services (GrabCar), • motorcycle taxis (GrabBike), • social carpooling (GrabHitch)	Ride-hailing (Other services: food, grocery and package delivery, and financial services)	Singapore, Malaysia, Cambodia, Indonesia, Myanmar, Philippines, Thailand and Vietnam	Drivers, i.e., individuals owning a vehicle
• EzyHaul	Road freight platform, cross-border shipments	Malaysia, Thailand, India and Singapore.	Transporters, carriers
GoGet	On demand work services	Malaysia	Part-timers with expertise in related task
PantangPlus	Traditional confinement services	Malaysia	traditional confinement therapists

Sources: Adapted from Constantiou et al. (2017) with authors' own analysis.

intermediaries, matchmakers, or gatekeepers for a fee. In doing so, they reduce risks, foster trust among participants, and lower transaction costs for their user bases (Constantiou et al., 2017). For instance, Grab connects people in need of a ride with drivers who are willing to drive them in their own cars. Grab handles transactions between customers and drivers and charges a fee for its services, which include screening drivers and their vehicles and running a peer-review system for riders and drivers. The use of advanced match-making algorithms reduces search costs by making it easier for drivers and passengers to connect with each other. Grab peer-reputation system, where drivers and passengers rate each other, lowers transaction barriers.

As Table 10.1 indicates, some of the major sectors of sharing economy include ride-hailing, accommodation, freelance work, and entertainment services. Airbnb and Uber are examples of service platform in the global level and Grab is a typical example in the regional level. The sharing economy involves four major archetypes, namely access to tangible assets (i.e., e-hailing and accommodation sharing), access to intangible services (freelance or part-time work, skills and talents), goods re-distribution (buy, sell and donation), investments and money transaction (i.e., investment and peer-to-peer lending).

Confluence of Digitalization and Shared Economy

The advancement of digital technology contributes to a rise of the new sharing economy, which could be called, the "digital sharing economy." In other words, the confluence of digitalization and sharing economy has expanded both resources and geographical aspect of traditional sharing economy. Regarding its resource's aspect, digital technology would enable people to share a wider choice of shareable resources. Regarding its geographic aspect, digital technology also would enable people to share things with a wider range of other people who may stay in distant geographic location (Pouri & Hilty, 2021). Younger generations, especially millennials who are born between the 1980s and the 2000s, have faithfully embraced this new trend of digital sharing economy. In other words, these younger people are more digital savvy consumers who are comfortable to use the shared goods and service, rather than to own them (Godelnik, 2017). Under the name of "collaborative consumption," the digital sharing economy becomes a part of lifestyle in new generations to create a new identity for them. In other words, the old identity in older generation was "you are what you own." By contrast, the new identity in younger generation could be expressed as "you are what you access" (Belk, 2014).

In this context, Jeremy Rifkin claimed that new trend of the sharing economy would bring forward a paradigm shift in the prevailing economic system by enabling people to realize great benefits to access the shared goods and service, rather than to own them. This kind of new consumption behaviour may contribute to improve a resource allocation by sharing the unused things to someone who would need them. He also argued that new idea of the digital sharing economy has already transformed some major industries, such as music, media, and education industries. In the music industry, numerous artists share their songs in internet sites without any charge. Some media firms post their articles in their web pages for free. Many prestige universities offer the free massive open online courses (MOOCs) for millions of students who are willing to register these courses. In other words, the power of digitalization and enormous potential of sharing economy would cause an eclipse of old capitalism system and a rise of new economic system which is based on the collaborative commons (Rifkin, 2014).

Among sharing economy platform firms, DiDi has a strong faith in a great potential of digital sharing economy and become a dominant ride-sharing firm in China with 550 million users and 30 million registered drivers. At the initial stage, DiDi followed Uber's business model that would create a match between drivers and passengers. Later, this platform firm saw a new business opportunity for the driving-licence holders who could not afford to own a car. The firm started offering a new matching service to share (lease) car to drivers who have a driving licence but have no car. This car-leasing service becomes an integrated component of their business model. As a result, DiDi significantly increase the number of their registered drivers (Jia et al., 2020).

Before the rise of digitalization, people shared a limited choice of shareable resources with a limited range of people who may stay in geographic close proximity, which could be called the "intimate sharing economy." In this traditional sharing economy, some resources are allocated as commons which are accessible to anyone within intimate social group. This kind of sharing economy is the most universal form of economic activities in any society (Price, 1975). In this context, economic activity could be understood as not only a market-based interaction between demand and supply under a constraint of the scarce resources, but a social provisioning process that would provide goods and services for a society (Gruchy, 1987; Jo, 2011). The concept of social provisioning process is similar with Polanyi's concept of the "provisioning" which spells out a role of market as a mean to allocate resources accordance of human needs, rather than a vehicle to maximize the individual utility (Brechin & Fenner, 2017; Polanyi, 1957). For example, the community-supported agriculture (CSA) could be a good example to highlight a social provisioning process. The CSA is a kind of subscription farming in which the members buy a share in a local farm, receive a portion of farm's output as a box of agriculture products and share excess foods with neighbours. This kind of traditional sharing economy would prevent a "tragedy" of commons which is over-exploitation of shared resources, such as community land, water, and air (Thompson & Press, 2014).

From a theoretical perspective, a sharing behaviour could be a key factor to offer some valuable insights on consumer behaviour in any society. However, this aspect of consumer behaviour is often confused with similar but different concepts, such as an act of reciprocation (Belk, 2010). For instance, the act of the sharing behaviour, such as food sharing, is seen as an act of reciprocation within a traditional society. This is mainly because a reciprocal obligation for giving something else is expected in returns of favour for receiving something. In this sense, basic motivation behind this sharing activity is to improve the efficiency of resource allocation by achieving economies of scale (Gurven, 2006). However, reciprocal obligation for return does not always accompany with the sharing behaviour in modern society. The sharing could be purely a redistributive practice which may not a part of reciprocation (Benkler, 2004). Based on Benkler's interpretation, Belk (2007) sees a sharing behaviour as a process to blur a difference between *mine* and *yours* by giving something to others and by receiving something else from others, without reciprocal obligation. In other words, the sharing behaviour would reformulate a private ownership by creating *ours*.

An interesting aspect in the process of creating a new form of ownership is that the shared goods and service have a double-excludability. By contrast, non-shared goods and services have only a single excludability. In other words, owners of non-shared goods and services would exclude those who are not willing to pay the market prices for their usage. This excludability is called as a market selection. However, shared goods and services tended to have another excludability which is known as a social section. It means that owners of shared good and service may exclude those who do not meet owner's social criteria (Benkler, 2004). The social section has a close linkage with issues of digital discrimination. The digital discrimination in the sharing economy could be defined as a situation in which some users are less favourably treated than other user on the base of some social criteria. For example, Airbnb created a platform that its hosts could choose guest on the base of some social criteria, such as ethnic group. In response to media criticism on this practice, the firm tried to clear up this negative image by requiring hosts to follow the anti-discrimination policy. However, Airbnb received backlash from hosts who claims their rights to choose who would enter their home (Cheng & Foley, 2018). This case clearly highlights a complexity of digital discrimination in the sharing economy. The anti-discrimination laws are used to eliminate the discrimination in context of the long-term contract in real estate market. However, these laws may not be directly applicable to the short-term contrast between hosts and guests in the sharing economy (Edelman et al., 2017).

More importantly, a widespread of internet usage in the digitalization would create a higher degree of anonymity in the online transactions than offline ones. This situation may contribute to eliminate discrimination against the disadvantage people. For prominent example in offline discrimination, Ayres and Siegelman (1995) pointed out that disadvantaged people tend to pay higher price for the purchase of new car than normal price. However, Morton et al. (2003) clearly demonstrated that there was not discrimination against the disadvantage people in the online purchase of new car. Similarly, there was no price discrimination in other online purchase through Amazon or eBay (Edelman et al., 2017). This means that digitalization may eliminate the price discrimination against disadvantaged people. More precisely, online transaction of non-shared goods and services tends to have a single excludability in which market selections would dominate its transaction. The owners of non-shared good and service in online transaction would made decision on the base of solely monetary calculation whether buyers are willing to pay the market price.

Put into a nutshell, the confluence of digitalization and sharing economy would create a higher degree of anonymity in the transaction between owners and buyer of the sharable resources which may contribute to eliminate a discrimination against disadvantage people in terms of market selection. However, the digitalization may not have a significant role to eliminate a digital discrimination against disadvantage people in terms of social selection which remain as a fundamental thorny issue in the digital sharing economy.

Consumer Behaviour during COVID-19 Pandemic

Literature review reveals a high emphasis on the COVID-19 impact and customer behaviours during the pandemic. Literature on the impact of COVID-19 was focused on the business disruption such supply chain (Meyer et al., 2021), business uncertainty during pandemic (Brown & Rocha, 2020), work patterns such as work from home or remote working (Galanti et al., 2021; Mustajab et al., 2020), and workers and employment such as immigrant workers (Wahab, 2020). Alongside these studies, the literature also offered interesting research concerning consumer behaviour changes (Cucchiarini et al., 2021) related to the occurrence of the COVID-19 pandemic such as changes in spending pattern (Vázquez-Martínez et al., 2021) especially on household spending (Kim et al., 2021) and macroeconomic expectations at a local level. Consumer behaviour has been altered as a result of COVID-19, and research shows that the pandemic has had an impact on the sharing economy. Through platform-driven applications and new business models, the sharing economy is bridging the gap between supply and demand, as well as filling employment gap during the pandemic.

According to Hossain (2021), there are three phases that have been broadly identified in the Covid-19 period: (a) reacting, such as through panic buying (Loxton et al., 2020), discretionary purchasing habits (Di Crosta et al., 2021), hoarding (Kuruppu & De Zoysa, 2020), rejecting, losing jobs (Roy et al., 2021), and practicing additional hygiene; (b) coping, such as through remote work (Vyas & Butakhieo, 2021), home quarantine (Alfawaz et al., 2021), stress (Avery et al., 2021), and social distance measures; and (c) long-term planning, such as restoring consumption, establishing a new social identity, and finding stable jobs.

People's behaviour in the event of a pandemic is likely to be influenced by their risk perceptions and is expected to last a long time. According to Cruz-Cárdenas et al. (2021), there are two distinct types of consumer behaviour: purchasing necessities and safety equipment and limiting time spent outside the home. As a result of the pandemic, activities such as dining out, hosting in-person conferences, and cruising were either halted or significantly reduced. Travel agencies and tour operators were also affected.

Transportation and accommodation sectors have been negatively affected by the COVID-19 lockdown, while the sharing economy as a whole has seen a surge in searches for services including freelance work and streaming services (Batool et al., 2021). Research on consumer behaviour in the context of a pandemic crisis has focused on both developed and developing countries in regard to the sharing economy. Ride-sharing and hospitality companies Airbnb and Uber are the focus of research in developed countries (Vinod & Sharma, 2021).

In a global economy where goods, services, people, and financial capital flow across borders, panic has distorted consumption patterns and the market system. The pandemic has fuelled an already-expanding trend of online shopping. It has encouraged experimentation by enticing consumers to try new ways to access products and services. Cashless payment methods have also grown in popularity as a result of social distancing measures. According to a recent Mastercard report, Malaysia has the highest rate of mobile wallet usage in Southeast Asia, at 40 percent, followed by the Philippines (36percent), Thailand (27percent), and Singapore (26percent).

Impact of Digitalization on Domestic Consumers

The sharing economy which benefitted from the development of ICT gave Malaysian consumers numerous sharing service options as a digital sharing economy. Interestingly, they adapted to the activity successfully. Whilst sharing economy services that have become part of the Malaysian public remain as significant as ever before and after the COVID-19 pandemic, the pandemic has also certainly affected the traditional ecosystem of sharing activities. This section shows the characteristics of the digital sharing economy in Malaysia through the presentation of some empirical data.

Figure 10.2 shows the contribution of the digital economy to the Malaysian national economy before and during the COVID-19 pandemic. This data comprise the gross value added of the ICT industry and e-commerce of the non-ICT industry. ICT is predominantly a service industry such as telecommunications services. Between 2018 and 2020, the ICT industry's gross value added, and e-commerce increased 1.6 percent and 2.5 percent, respectively. Overall, the digital economy's contribution to the national economy recorded an increase of 4.1 percent, from 18.5 percent in 2018 to 22.6 percent in 2020.

Figure 10.3 shows the contribution of e-commerce, including the ICT industry and non-ICT industry, to GDP. Prior to the COVID-19 pandemic, there was only a slight overall increase in 2018–2019. In 2020 (COVID-19 time), it could be clearly seen there was an increase of 0.9 percent (ICT industry) and 2.4 percent (non-ICT industry) compared to the previous year, for a total increase of 3 percent. These data can be read as an indication of the importance of the digital economy in the country and the fact that the ICT-based sharing economy industry makes a certain contribution to the national economy. As can be seen in Table 10.2, the contribution of the e-commerce sector is also reflected in the amount of GDP: the GDP increase of the e-commerce sector from 2018 to 2019 was 13.7 billion MYR, whilst 34.1 billion MYR increase was observed from 2019 to 2020.

The Department of Statistics Malaysia (2019, 2021) provided practical point of view in terms of the influence and impact of domestic digital economy. Malaysia under the pandemic had 3 million new digital consumers by 2020 and the first half

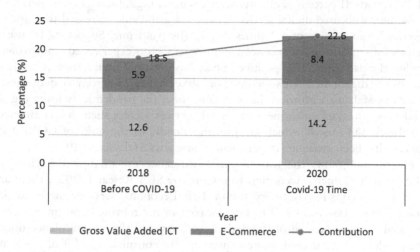

Figure 10.2 Contribution of Digital Economy, Malaysia, 2018–2020.

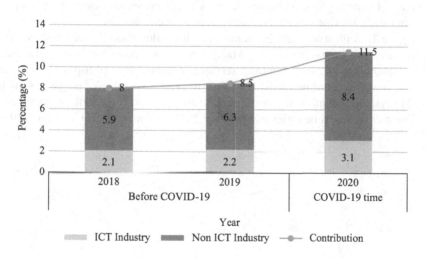

Figure 10.3 Contribution of E-Commerce to GDP, Malaysia, 2018–2020.

Table 10.2 E-Commerce Contribution to GDP, Malaysia, 2018–2020

GDP (Billion MYR)	115.5	129.2	163.3
Year	2018	2019	2020

Source: Department Statistics of Malaysia.

of 2021, with 81 percent of all internet users using digital services. Furthermore, consumers who used digital services before the pandemic increased their use of services by an average of 4.2 times during the pandemic. Supported by users' 76 percent satisfaction with services, all internet sectors experienced rapid growth during the pandemic. This positive impact has not only been felt by consumers but also by digital merchants and investors alike. Whilst 43 percent of digital merchants in Malaysia withstood the recession under the pandemic by using digital platforms, investors' investment in digital services sectors such as e-commerce, healthtech and fintech, which are growing rapidly in the wake of COVID-19 appetite has been growing (Department of Statistics Malaysia, 2019, 2021).

Figure 10.4 shows the reasons for continuance use of digital services on Malaysian consumers. According to e-Conomy SEA Research 2021, there are main two reasons for continued use of digital economy services: made my life easier/ more convenient and to became part of my routine. Consumer trends revealed that the growth of the digital economy and the inconvenience under the pandemic have made it more convenient for consumers and it has become part of their lives. Consistent across the four e-commerce categories is a high degree of convenience and simplification of consumers' lives. As for the ICT sector, consumers feel food delivery services particularly convenient. The point that may be highlighted is that music and video services have been already a part of their daily routine as well as a convenience for consumers.

Figure 10.5 illustrates the gross merchandise value (GMV), which shows the total value of purchases made by Malaysian customers across four digital platforms including sharing economy services: e-commerce, transport and food, online travel, and online media. GMV shows the total value of purchases made by Malaysian customers in e-commerce, transport and food, online travel, and online media. The time series is as follows: 2019 (before COVID-19 time), 2020

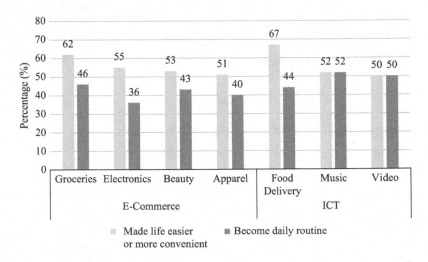

Figure 10.4 Reasons for Malaysian Consumers' Continued use of Digital Services, 2021.

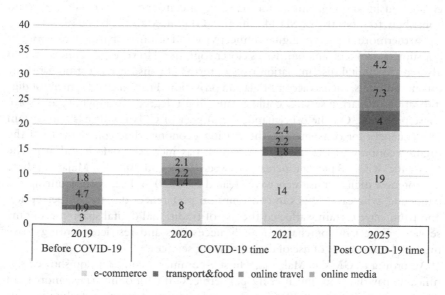

Figure 10.5 GMV per Sector, Malaysia, 2019–2025 (US$ Billion).

and 2021 (COVID-19 time), and 2025 (post COVID-19 time). Throughout these three stages, customer usage in the digital economy is shown to continue to increase and especially e-commerce plays the vital role to attract consumers to the digital economy. In 2020–2021, there was a sharp rise of USD 6 billion in GMV. Regarding transport & food, including ride-hailing and online food delivery (OFD), ride-hailing has been affected by the pandemic, but the market is expected to grow due to increased consumers' demand for OFD. By 2025, the market is expected to reach a GMV of USD 4 billion. The online travel sector, affected by strict MCOs and lockdowns by the Malaysian government, saw more than double the GMV phenomenon compared to COVID-19. On the other hand, the sector is expected to recover and grow after the pandemic. Netflix and Amazon online media streaming services such as Prime saw GMV growth of USD 0.6 billion from 2019 to 2021 due to the growth of the industry itself before the pandemic and the outage restrictions under the pandemic. The data estimated that GMV will increase to USD 4.2 billion by 2025.

Discussion and Implications

Shared economy activities benefitting from the digital economy have had a significant impact on consumers' own digitalization. In particular, the digitalization of the sharing economy has expanded the scope of resource geographical adaptation and created a new identity of communal consumption (digital sharing economy) among consumers, especially the younger generation. This paradigm shift in the existing economic system, led by digital platforms, has created business opportunities for many co-operates and individuals and could contribute to more than USD 14 billion in GDP in Malaysia by 2025.

Furthermore, aspects of digital connectivity and resource sharing have brought consumers, workers, and businesses closer together. However, this phenomenon also creates digital discrimination risks in terms of consumer selection for some business models, such as accommodation provision. Further development of digitalization in sharing services could create a high degree of anonymity and solve these problems. On the other hand, the impact of COVID-19 has had certain consequences for consumers in the sharing economy. The contribution of the digital economy to the domestic economy has increased before and after the pandemic, with a 3 percent increase between 2019 and 2020. In Malaysia, three million new digital consumers were created by 2020 and early 2021 during the pandemic, whereas changes in consumption preferences due to risk perception and panic have certainly affected the use of traditional digital sharing economy services. Especially, purchases of daily necessities and restricted outings have impacted many areas of use of digital sharing services.

According to GMV in Malaysia, whilst e-commerce consumption showed significant growth before and after the pandemic, GMV for online travel more than doubled from 2019 to 2020. Moreover, ride-hailing has seen a decline in consumer demand due to curfew and fewer drivers, whilst the increase in demand for online food delivery led to an increase in overall GMV. Ride-hailing giants such as Grab and Uber that have a hybrid business model that encompasses elements of the gig economy in a sharing economy concept have been supported by online food delivery revenues even during the pandemic. Consumers perceived increased convenience and daily use under the COVID-19 pandemic as factors for continued use of these services.

As a result, COVID-19 has brought about certain changes in the traditional digital sharing economy consumption system in Malaysia, with consumers beginning to identify the use of these services as a daily routine, as the convenience of digital technology coexists with the inconvenience under the pandemic. There is growth potential in the convergence of digitalization and the sharing economy and its collaboration with consumers, given that the estimation of increase of GMV of the main digital sharing economy sectors in 2025. At the same time, the future research needs to focus on how digital technologies can compensate for the risks that occur.

Conclusion

There are several gaps in our knowledge in research as a result of our findings, so there are numerous research opportunities available. Researchers will be able to propose a new conceptual framework in future study in order to further investigate consumer preferences in the sharing economy during the COVID-19 pandemic. Practically, to ensure that the former benefit fairly from their efforts, guidelines for SE service providers and receivers must be established. It is possible that future study will reveal ways to increase workers' economic stability, employment support, and the introduction of social benefits in the SE sector. Although the COVID-19 stresses the adoption of new norms emphasising social distance, hygiene, and security, how this relates to the SE needs further exploration. COVID-19 has had some beneficial results in that it has compelled numerous stakeholders to improve their sharing services and build them into an established sector, and future research may study various ways to do this. Scholars may be able to recommend tax policies for SE firms because, in contrast to traditional counterparts like taxi services, SE firms do not pay their fair share of taxes. As a result of the COVID-19's impact on the SE, researchers have been able to examine ways in which this industry might grow despite the current crisis.

Acknowledgement

Authors gratefully acknowledge financial support from the Fundamental Research Grant Scheme (FRGS) from Ministry of Higher Education, Malaysia (Project No. FP026–2021: Project code: FRGS/1/2021/SS0/UM/02/12).

References

Agarwal, N., & Steinmetz, R. (2019). Sharing Economy: A Systematic Literature Review. *International Journal of Innovation and Technology Management, 16*(6), 1–17. https://doi.org/10.1142/S0219877019300027

Alfawaz, H., Amer, O. E., Aljumah, A. A., Aldisi, D. A., Enani, M. A., Aljohani, N. J., Alotaibi, N. H., Alshingetti, N., Alomar, S. Y., Khattak, M. N. K., Sabico, S., & Al-Daghri, N. M. (2021). Effects of Home Quarantine during COVID-19 Lockdown on Physical Activity and Dietary Habits of Adults in Saudi Arabia. *Scientific Reports, 11*(1), 1–7. https://doi.org/10.1038/s41598-021-85330-2

Avery, A. R., Tsang, S., Seto, E. Y. W., & Duncan, G. E. (2021). Differences in Stress and Anxiety among Women With and Without Children in the Household During the Early Months of the COVID-19 Pandemic. *Frontiers in Public Health, 9*(September), 1–14. https://doi.org/10.3389/fpubh.2021.688462

Ayres, I., & Siegelman, P. (1995). Race and Gender Discrimination in Bargaining for a New Car. *The American Economic Review, 85*, 304–321.

Barbu, C. M., Florea, D. L., Ogarcă, R. F., & Răzvan Barbu, M. C. (2018). From Ownership to Access: How the Sharing Economy Is Changing the Consumer Behavior. *Amfiteatru Economic, 20*(48), 373–387. https://doi.org/10.24818/EA/2018/48/373

Bardhi, F., & Eckhardt, G. M. (2012). Access-Based Consumption: The Case of Sharing. *Journal of Consumer Research, 39*(December), 881–898.

Batool, M., Ghulam, H., Hayat, M. A., Naeem, M. Z., Ejaz, A., Imran, Z. A., Spulbar, C., Birau, R., & Gorun, T. H. (2021). How COVID-19 Has Shaken the Sharing Economy? An Analysis Using Google Trends Data. *Economic Research-Ekonomska Istraživanja, 34*(1), 2374–2386. https://doi.org/10.1080/1331677X.2020.1863830

Belk, R. (2007). Why Not Share Rather Than Own? *Annals of the American Academy of Political and Social Science, 611*, 126–140.

Belk, R. (2010). Sharing. *Journal of Consumer Research, 36*(5), 715–734.

Belk, R. (2014). You Are What You Can Access: Sharing and Collaborative Consumption Online. *Journal of Business Research, 67*(8), 1595–1600.

Benkler, Y. (2004). Sharing Nicely: On Shareable Goods and the Emergence of Sharing as a Modality of Economic Production. *Yale Law Journal, 114*, 273–358.

Brechin, S. R., & Fenner, W. H. (2017). Karl Polanyi's Environmental Sociology: A Primer. *Environmental Sociology, 3*(4), 404–413.

Brown, R., & Rocha, A. (2020). Entrepreneurial Uncertainty During the Covid-19 Crisis: Mapping the Temporal Dynamics of Entrepreneurial Finance. *Journal of Business Venturing Insights, 14*, 1-10.

Cheng, M. (2016). Sharing Economy: A Review and Agenda for Future Research. *International Journal of Hospitality Management, 57*, 60–70. https://doi.org/10.1016/j.ijhm.2016.06.003

Cheng, M., & Foley, C. (2018). The Sharing Economy and Digital Discrimination: The Case of Airbnb. *International Journal of Hospitality Management, 70*, 95–98.

Constantiou, I., Marton, A., & Tuunainen, V. K. (2017). Four Models of Sharing Economy Platforms. *MIS Quarterly Executive, 16*(4), 236–251.

Cruz-Cárdenas, J., Zabelina, E., Guadalupe-Lanas, J., Palacio-Fierro, A., & Ramos-Galarza, C. (2021). COVID-19, Consumer Behavior, Technology, and Society: A Literature Review and Bibliometric Analysis. *Technological Forecasting and Social Change, 173*, 1-13 https://doi.org/10.1016/j.techfore.2021.121179

Cucchiarini, V., Caravona, L., Macchi, L., Perlino, F. L., & Viale, R. (2021). Behavioral Changes after the COVID-19 Lockdown in Italy. *Frontiers in Psychology, 12*(March), 1–24. https://doi.org/10.3389/fpsyg.2021.617315

Di Crosta, A., Ceccato, I., Marchetti, D., la Malva, P., Maiella, R., Cannito, L., Cipi, M., Mammarella, N., Palumbo, R., Verrocchio, M. C., Palumbo, R., & Domenico, A. Di. (2021). Psychological Factors and Consumer Behavior during the COVID-19 Pandemic. *PLoS ONE, 16*(8 August), 1–23. https://doi.org/10.1371/journal.pone.0256095

Department of Statistics Malaysia (2019, October 16). *Contribution of Digital Economy Was 18.5 Per cent to National Economy.* https://www.dosm.gov.my/v1/index.php?r=column/cthemeByCat&cat=473&bul_id=UERrcjExbnRKd3NwQ0Y3V011RngyQT09&menu_id=b0pIV1E3RW40VWRTUkZocEhyZ1pLUT09

Department of Statistics Malaysia (2021, November 10). *Malaysia E-commerce Income Soared 17.1 Per cent to RM279.0 Billion in the Third Quarter 2021.* https://www.dosm.gov.my/v1/index.php?r=column/cthemeByCat&cat=473&bul_id=cmRYZ-21sUVF4elBySHVWckhkMGU4Zz09&menu_id=b0pIV1E3RW40VWRTUkZocEhyZ1pLUT09

Economic Planning Unit Prime Minister's department (n.d.). *Malaysia Digital Economy Blueprint.* https://www.epu.gov.my/sites/default/files/2021-02/malaysia-digital-economy-blueprint.pdf

Edelman, B., Luca, M., & Svirsky, D. (2017). Racial Discrimination in the Sharing Economy: Evidence from a Field Experiment. *American Economic Journal: Applied Economics, 9*, 1–22.

Galanti, T., Guidetti, G., Mazzei, E., Zappalà, S., & Toscano, F. (2021). Work From Home During the COVID-19 Outbreak: The Impact on Employees' Remote Work Productivity, Engagement, and Stress. *Journal of Occupational and Environmental Medicine, 63*(7), E426–E432. https://doi.org/10.1097/JOM.0000000000002236

Godelnik, R. (2017). Millennials and the Sharing Economy: Lessons from a 'Buy Nothing New, Share Everything Month' project. *Environmental Innovation and Societal Transitions, 23*, 40–52.

Gruchy, A. G. (1987). *The Reconstruction of Economics: An Analysis of the Fundamentals of Institutional Economics.* New York: Praeger.

Gurven, M. (2006). The Evolution of Contingent Cooperation. *Current Anthropology*, 47(1), 185–192.

Hamari, J., Sjöklint, M., & Ukkonen, A. (2016). The Sharing Economy: Why People Participate in Collaborative Consumption. *Journal of the American Society for Information Science and Technology*, 67, 2047–2059. https://doi.org/10.1002/asi

Hossain, M. (2021). The Effect of the Covid-19 on Sharing Economy Activities. *Journal of Cleaner Production*, 280, 1-9

Rifkin, J. (2014). *The Zero Marginal Cost Society: The Internet of Things, the Collaborative Commons, and the Eclipse of Capitalism*. New York: St. Martin's Press.

Jia, F., Li, D., Liu, G., Sun, H., & Hernandez, J. E. (2020). Achieving Loyalty for Sharing Economy Platforms: An Expectation–Confirmation Perspective. *International Journal of Operations & Production Management*. 40, 1067–1094.

Jo, T. H. (2011). Social Provisioning Process and Socio-Economic Modelling. *American Journal of Economics and Sociology*, 70(5), 1094–1116.

Kana, G. (2021). A Missed Opportunity for Malaysia. *The Star* dated 15 April, 2021 https://www.thestar.com.my/business/business-news/2021/04/15/grab---a-missed-opportunity-for-malaysia [accessed on 2 April, 2022]

Kim, S., Koh, K., & Zhang, X. (2021). Short-term Impact of COVID-19 on Consumption Spending and its Underlying Mechanisms: Evidence from Singapore. *Canadian Journal of Economics*, 0(0), 1–20. https://doi.org/10.1111/caje.12538

Kuruppu, Gayithri Niluka and De Zoysa, Anura, COVID-19 and Panic Buying: An Examination of the Impact of Behavioural Biases. Available at SSRN: https://ssrn.com/abstract=3596101 or HYPERLINK "https://dx.doi.org/10.2139/ssrn.3596101" http://dx.doi.org/10.2139/ssrn.3596101

Loxton, M., Truskett, R., Scarf, B., Sindone, L., Baldry, G., & Zhao, Y. (2020). Consumer Behaviour During Crises: Preliminary Research on How Coronavirus has Manifested Consumer Panic Buying, Herd Mentality, Changing Discretionary Spending and the Role of the Media in Influencing Behaviour. *Journal of Risk and Financial Management*, 13(166), 1-21

Luri Minami, A., Ramos, C., & Bruscato Bortoluzzo, A. (2021). Sharing Economy Versus Collaborative Consumption: What Drives Consumers in the New Forms of Exchange? *Journal of Business Research*, 128(February), 124–137. https://doi.org/10.1016/j.jbusres.2021.01.035

Meyer, A., Walter, W., & Seuring, S. (2021). The Impact of the Coronavirus Pandemic on Supply Chains and Their Sustainability: A Text Mining Approach. *Frontiers in Sustainability*, 2(March), 1–23. https://doi.org/10.3389/frsus.2021.631182

Mustajab, D., Bauw, A., Rasyid, A., Irawan, A., Akbar, M. A., & Hamid, M. A. (2020). Working From Home Phenomenon as an Effort to Prevent COVID-19 Attacks and its Impacts on Work Productivity. *The International Journal of Applied Business*, 4(1), 13. https://doi.org/10.20473/tijab.v4.i1.2020.13-21

Morton, F. S., Zettelmeyer, F. & Silva-Risso, J. (2003). Consumer Information and Discrimination: Does the Internet Affect the Pricing of New Cars to Women and Minorities? *Quantitative Marketing and Economics*, 1, 65–92.

Oi, M. (2021). *Ride Hailing App Grab Falls in $40bn Market Debut*. BBC dated 2 December, 2021. https://www.bbc.com/news/business-59486675 [accessed on 2 April 2022].

Polanyi, K. (1957). *The Great Transformation*. New York: Rinehart.

Pouri, M. J., & Hilty, L. M. (2021). The Digital Sharing Economy: A Confluence of Technical and Social Sharing. *Environmental Innovation and Societal Transitions*, 38, 127–139.

Price, J. A. (1975). Sharing: The Integration of Intimate Economies. *Anthropologica*, 3–27.

Roy, S., Dutta, R., & Ghosh, P. (2021). Identifying Key Indicators of Job Loss Trends During COVID-19 and Beyond. *Social Sciences & Humanities Open*, 4(1), 100163. https://doi.org/10.1016/j.ssaho.2021.100163

Thompson, C. J., & Press, M. (2014). How Community-Supported Agriculture Facilitates Re-embedding and Reterritorializing Practices of Sustainable Consumption. In J. Schor, & C. J. Thompson (Eds.), *Sustainable lifestyles and the quest for plenitude: Case studies of the new economy* (pp. 125-147). Yale University Press.

Vázquez-Martínez, U. J., Morales-Mediano, J., & Leal-Rodríguez, A. L. (2021). The Impact of the COVID-19 Crisis on Consumer Purchasing Motivation and Behavior. *European Research on Management and Business Economics*, 27(3), 100166. https://doi.org/10.1016/j.iedeen.2021.100166

Vinod, P. P., & Sharma, D. (2021). COVID-19 Impact on the Sharing Economy Post-Pandemic. *Australasian Accounting, Business and Finance Journal*, 15(1 Special Issue), 37–50. https://doi.org/10.14453/aabfj.v15i1.4

Vyas, L., & Butakhieo, N. (2021). The Impact of Working From Home During COVID-19 on Work and Life Domains: An Exploratory Study on Hong Kong. *Policy Design and Practice*, 4(1), 59–76. https://doi.org/10.1080/25741292.2020.1863560

Wahab, A. (2020). The Outbreak of Covid-19 in Malaysia: Pushing Migrant Workers at the Margin. *Social Sciences & Humanities Open*, 2(1), 100073. https://doi.org/10.1016/j.ssaho.2020.100073

Wirtz, J., So, K. K. F., Mody, M. A., Liu, S. Q., & Chun, H. E. H. (2019). Platforms in the Peer-to-Peer Sharing Economy. *Journal of Service Management*, 30(4), 452–483. https://doi.org/10.1108/JOSM-11-2018-0369

11 The Role of TVET in IR4.0 for Malaysia

Cheong Kee Cheok and Li Ran

Introduction

The advance of Industrial Revolution 4.0 (IR4.0) suggests either its inevitability or that its overwhelming advantages would persuade industry to embrace this new technology. But seldom are its costs highlighted. These include the social disruptions that adoption of the technology brings. This include magnifying the divides between the haves (with Internet) and the have-nots (without), polarization of jobs leading likewise to greater inequality, greater capital intensity and skilled labour shortages, all contributing to greater welfare vulnerability for some population groups (Sony, 2020: 260).

But what precisely is IR4.0? The Industrial Revolution began with the invention of steam-powered engines for industry and transportation transformed both from the late 18th century. The Second Industrial Revolution was said to occur in the early 20th century with the arrival of electric power, which became the driver of both industry and home. The Third Industrial Revolution was associated with the emergence of ICT in industry and the development of personal computers, these developments beginning in the 1970s. The beginning of 2000s saw the beginning of the Fourth Industrial Revolution (IR4.0) also referred to as the "digital revolution" which combines human and technological capabilities in industry, beginning with the most technologically advanced countries.

Assuming arguments for IR4.0 are favourable, its onset prompts each country to ask itself at least two questions. First, what is the country's level of attainment of IR4.0? Put another way, how far short is the country's attainment of IR4.0 that education, in whatever form, must bridge? A whole series of sub-questions can be asked as to what precipitated this gap. Or whether there exist factors that inhibit bridging of the gap. And how can these challenges be overcome?

Second, in preparing the population to take advantage of IR4.0, or at least to bring it up to speed with being part of the IR4.0 ecosystem, what is the role of education and what challenges does this sector or subsector face in pursuit of this goal? What does education for IR4.0 entail? What stumbling blocks exist to impede its progress? Are there subsectors of education that are particularly suited for nurturing IR4.0 upskilling? To the extent that the country has not reached the level to be IR4.0 compliant, is it necessary for education to go all the way?

DOI: 10.4324/9781003367093-11

To these core questions can be added a third. Given the pivotal role of the government in promoting IR4.0, whether by adopting the technology needed for IR4.0 implementation, or by management of the education system that produces ultimately IR4.0-capable graduates. Or fail to do so in one or both cases. What accounts for its success in delivering on both scores, or what constraints are faced in the pursuit of either or both objectives. Was there adequate effort or was there consonance between planning and implementation?

These two questions are what this chapter attempts to answer. In doing so, it first sketches the meaning and significance of IR4.0. Also highlighted are extant government strategies to promote IR4.0 and the use of technical and vocational education and training TVET as an instrument to familiarize the population. Examples are also drawn from other countries' experiences.

We pose these questions in the case of Malaysia, a high middle-income country poised to advance to developed country status in 2020, according to then Prime Minister Mahathir's Wawasan (Vision) 2020 blueprint[1], but on account of the country's slower growth since, was pushed back to year 2030 by Prime Minister Najib Razak. Even at this lengthened schedule, doubts have been expressed as to whether this schedule is achievable, given the challenges like deindustrialization (Rasiah, 1995, 2011, 2020) and politics (Gomez et al., 2021) the country has to confront. To achieve its goal of sustained growth, mastering IR4.0 to leverage the latest technology is essential. How far is Malaysian industry from the technological frontier?

Malaysia in the IR4.0 Era

This chapter has the specific purpose of ascertaining the role of TVET in upgrading the requisite skills to take advantage of the bounty of IR4.0. TVET is chosen over academic education because RVET is specifically about vocational training. To ascertain the readiness of TVET for readying citizens for the onset of IR4.0, two related issues must be addressed. First, even before the efficacy of TEVT is discussed, how prepared is Malaysia to master the technologies associated with IR4.0? Second, given the technologies expected in IR4.0, what role can TVET play in upgrading skills. We examine the first question first.

Malaysia's Preparedness for IR4.0

There is no question of the government's awareness of the arrival of IR4.0. In 2021, no fewer than three documents referenced policies or activities related to IR4.0. In February 2021, the government released *Malaysia Digital Economy Blueprint* (MDEB) (EPU, 2021a). The MDEB was intended to support other government initiatives, but specifically charts the course of MYDIGITAL, "a national initiative ... of the Government to ... transform Malaysia into a digitally-driven, high income nation and a regional leader in digital economy" (EPU, 2021a: 10). Its Phase 1 (2021 to 2022) aims to accelerate adoption of the digital foundation needed for Phase 2 (2023–2025) which aims to drive digital transformation, and Phase 3 (2026–2030) which builds strong, sustainable growth in the decades to come, positioning Malaysia to become a regional market producer for digital products and digital solutions provider (EPU, 2021a: 12). The digital economy is of course a part of IR4.0.

In July 2021, the Malaysian government announced through its *National Fourth Industrial Revolution (IR4) Policy* (Malaysia (2018) its policy on IR4.0, a "broad overarching national policy that drives ... the adoption of emerging technologies ... and builds the foundation to drive digitalization across the nation, including bridging the digital gap" (EPU, 2021a: 27). With policy thrusts that included upgrading the citizens' IR4.0 skills to cope with IR4.0 and accelerate IR4.0 innovation and adoption (EPU, 2021b: 33). For all three stakeholder groups – businesses, society and government, it would focus on ten key sectors and six supporting sectors. Confidence in achieving its targets was based on Malaysia's readiness to embrace technology through plans like the National Policy on Science, Technology and Innovation 2013–2020, National Education Blueprint 2013–2025, National Internet of Things (IoT) Strategic Roadmap 2015–2025, Eleventh Malaysia Plan 2016–2020 (11MP) with strategy to expand modern services, Malaysia Productivity Blueprint. And Malaysia's "above average ranking in key technology and innovation related global indices ... (which) has provided a strong foundation for the nation to seize growth opportunities and mitigate risks arising from the IR4.0" (EPU, 2021b: 33). This is despite Malaysia's admitted decline in competitiveness as shown by global competitiveness indices.

In September 2021, the Malaysian government released the Twelfth Malaysia Plan 2021–2025. Though not targeted specifically at IR4.0, it made ample references to it, beginning with accelerating technology adoption and innovation as a "policy enabler" and "improving the TVET ecosystem to produce future-ready talent." In giving ample coverage of IR4.0 and the digital economy, the Plan is explicit on the challenges faced, including low investment in R&D, low commercialization and experimental research and shortage of STEM graduates. The declining trend in Malaysia's competitiveness and product innovation has also been noted Bernama, 2019.

With these initiatives in place but given the acknowledged challenges, how do the optimistic projections compare with observations on the ground? Although piecemeal, the picture that emerged is mixed – major deficiencies are reported as frequently as stories of successful implementation. In a reported interview on July 28, 2017 (Mibrand, 2017), the chief executive officer of the Human Resources Development Board (HRDB), Vinaeswaran Jeyandran was of the view that despite the hype, Malaysia is not yet in IR4.0. Just over 30 percent of Malaysia's 15 million workforce has been equipped to handle IR4.0 processes. HRDF has its work cut out for it to increase the proportion to 35–40 percent. *Techwire Asia* reports an even lower adoption among Malaysian SMEs of 10 to 15 percent taking steps to adopt IR4.0 processes with the reasons cited being the lack of structure and collaboration between the private and public sectors in driving the country's IR4.0 agenda (Techwire Asia, 2019). [2] Uncertainty as to how to access government funds to migrate to IR4.0 proved to be an additional hurdle.

A separate study found fault with Malaysian Company research innovations which focused on unlocking value in new businesses rather than revitalize their core businesses (Techwire Asia, 2019).

No more optimistic was Kaur (2019) who concluded that Malaysia was falling behind in IR4.0 adoption. *TM One* (2019) concurred.

> In general, Malaysia is still struggling to adopt IR4.0, and many businesses are stuck at Industry 3.0, in terms of manufacturing technology... many manufacturers still rely on low-cost labour, i.e., foreign workers, and are hesitant to invest in innovative automation technologies.

Among other manufacturers, those that adopted IR4.0 were no fully cognizant of its impact (Nasir, 2019) while those that had not adopted did not what it was for (Lee et al., 2019).

But surveys of members of civil society offer no more clarity. Idris' (2019) survey of 400 public university students found a majority who had heard of IR4.0 (90 percent), had some understanding of IR4.0 (70 percent), recognized the importance of education funding to upgrade IR4.0 knowledge and skills (75 percent), were aware of the advantages of IR4.0 in terms of lowering production cost (65 percent), reduced manual work (65 percent), increased income (50 percent) and increase employment (30 percent). They also felt that universities effort in promoting IR4.0 was insufficient (55 percent) and that there was insufficient supporting infrastructure (60 percent). The opposite results came from another survey of students of a private university (Chalil, 2019; Sani, 2020). The survey of 550 student respondents found students (and parents) lacked clarity about and felt unprepared to join a IR4.0-capable workforce. They also felt that higher education was not doing enough. More than 50 percent of the students were unable to articulate about the substance of IR4.0, more than 50 percent of the parents were unable to discuss IR4.0. Respondents felt that universities stressed theoretical and academic teaching, and did not expose students to IR4.0. There was a need for soft skills – critical thinking, problem solving, leadership.

Beyond issues on the ground Malaysia encounters structural advantages as well as challenges. Lee (2020) summarized these well. Malaysia's advantages included:

- Its strategic location as an e-commerce and logistic hub, with multinationals like Nestle, Ikea, Lazada and Continental Tyres locating here.
- Its good infrastructure, with 30 highways, five international airports and seven international seaports having good access to all parts of the country, with the Malaysian Communications and Multimedia Commission 2018 statistics showing 39.4 broadband subscriptions, 3G and 4G/LTE systems reaching 94.7 and 79.7 percent population coverage.
- One of lowest rents in Grade A office space, the lowest living cost in Asia. Investment funds are also abundant, and developers/construction companies experienced.
- Industrial zones offering investment incentives

But these incentives are not unique to Malaysia. On the other hand, some challenges are:

- Companies reluctant to embrace change in the face of costs and uncertainty;
- Shortage of talent – of STEM graduates, and insufficient focus on digital literacy and innovation, resulting in a less adaptable workforce;
- Stringent regulation – red tape, endemic corruption, many sectors tightly controlled by state monopolies or GLCs; and
- TVET's limited success, graduate unemployment, racial discrimination and political issues prompting brain drain that adversely affected high-tech and high-value industrial production.

Add to these the phenomenon of premature deindustrialization (Rasiah, 2020)? and the plunge into real estate services (Gomez et al., 2021), it is difficult to concur with the government's rosy projections of accomplishments. Or doubt the various stakeholders' scepticism that IR4.0 is now in place. It is therefore little wonder that an article in the magazine *TMOne* (2019) asked: "Industry 4.0 is here: where are the manufacturers?" pointing its fingers at reliance on "low-skill foreign workers". Kaur (2019) also pointed out that only 15–20 percent of companies had migrated to IR4.0, the rest lacking the knowledge or information on how to borrow to upgrade. She also quoted consultancy International Data Corporation as saying that the workforce was similarly unprepared.

Higher Education and IR4.0

The arrival of IR4.0 poses major challenges for education, not least having to deal with so many changes in the environment in its role in nurturing the workforce to be IR4.0-ready. These changes include the nature of employment as there could be reduced demand for entry-level graduates while those with specializations would remain to be sought after. The structure of employment will also change as the demand for different types of goods and services change. Technologies at work and for learning are also changing. Partly responding to technological changes, student study habits will evolve to take advantage of new capabilities. The new environment will require major restructuring of education systems and pedagogy just to keep pace. At the same time these disruptions offer a real opportunity for fundamental education reform that, if successful, will produce a creative and innovative workforce.

Given these deficiencies, what role can education, specifically higher education play to enable all stakeholders to be current in IR4.0 knowledge and use? And what type of education is best suited to ensuring proficiency? To answer these questions, those who will form the workforce in an IR4.0 environment need to be profiled. Kozinski (2017) characterized these Generation-Z students as fully engaged in learning but want to control this process. They do not fear challenges and enjoy group discussion and collaborate in teams in a highly interactive environment. For them, learning is not limited geographically and intertemporally; they can learn anywhere and anytime and have familiarity with the latest technology.

In this environment, what constitutes the appropriate pedagogy and approach that can be effectively utilized? Haseeb (2018) provided some possibilities. Stressing that "Education must produce creative graduates, capable of critical thinking, innovative and entrepreneurial, have cognitive flexibility to deal with complexity", Haseeb (2018) was of the view that pedagogically, communication and collaborating and collaborative skills would assume greater prominence, together with digital and data literacy. There should be opportunities for more individualized modes of learning to suit individual students' needs. He also foresaw blended learning between traditional instruction and massive open online courses (MOOCs) and deeper learning through more use of practice-oriented learning or learning-by-doing seeing more frequent use. More and quicker disruptions would require life-long learning plus shorter duration of programs. At the same time, the multitude of skills required in IR4.0 would mean the offer of packages of interdisciplinary course offerings such engineering, business administration and computer science. Corresponding to IR4.0, this state of education has been called Education 4.0.

The success of this transformation depends crucially on at least three factors. First, because digital technology relies on the internet, the use of this technology is available only to those with internet connections, possibly magnifying the divide between the more affluent haves and the less affluent have-nots. Second, even for students with the means to develop new ways of learning, the role of teachers is even more vital. Miranda et al. (2021: 1) listed the following attributes teachers needed to possess: competencies, learning methods, ICT and infrastructure. Even as students need to master vital media for guiding and transferring the new skills, teachers' commitment matters. Some resistance to change, given uncertainties over methods and outcomes, is inevitable. This resistance may take the form of complaints of insufficient time to integrate new content with old, focus on ensuring on students passing examinations, loss of control in trying new content, reliance on textbooks and lack of ICT familiarity. Arguably the greatest change in teacher mindset will come from recognizing that the student, rather than the teacher, should be the centre of attention (Gerstein, 2014).

TVET in Malaysia

TVET Evolution

The history of TVET in Malaysia can be traced to as early as 1964 when both ministries of Human Resources and Youth and Sports first provided skills training with more ministries setting up training institutions since (Amir, 2012). In 2013, a comprehensive review of Malaysia's TVET at that time was conducted under the auspices of the World Bank's SABER (Systems Approach for Better Education Results) project to assess its strategic framework, system oversight and service delivery (Cheong et al, 2013). While documenting major improvements in

Malaysia's TVET system between year 2000 and 2010, it nevertheless noted areas where there existed room for improvement. Critiques related to first the greater attention paid to policy pronouncements than to implementation (Cheong et al., 2013: 15–16). Second, performance was measured by the allocations spent rather than by indicators of performance. Third, "the public sector focus of … programs afford only limited roles for non-government stakeholders." Fourth, public sector focus had been reduced by non-state design of programs, and industrial attachments but another group of stakeholders, the workers, remained unrepresented. Fifth, Malaysia had organizationally multiple public agencies entrusted with workforce training through TVET with limited coordination among them, although the harmonization of standards had ensured some consistency. Institutional issues like reporting, data dissemination, and assessment were less well coordinated. Sixth, the geographic distribution of (MARA) training centres raised the possibility of program duplication. Seventh, problems associated with the lack of institutional memory, caused partly by institutional restructuring but equally likely by the government's data confidentiality, had consequences for charting the longer-term development of Malaysia's TVET. On top of all these issues, the general public's perception of TVET being the refuge of those unable to make the grade academically, was not helpful in bringing students to TVET.

In the same year, Ismail and Norhasni (2014) added to this critique. Their additional criticisms are given in Table 11.1. In addition to improvements in the strategic framework, system oversight and service delivery, their criticisms centred around structural deficiencies like the shortage of qualified teaching staff, courses being supply-driven, and concentration in lower-level skills.

Nearly a decade has elapsed since this research, and significant changes have occurred since then. First, the 11MP was issued in 2016 which brought TVET into the education mainstream and gave the sub-discipline a major policy push. This is followed by the Twelfth Malaysia Plan that accorded even more publicity and generated more discussion of mainstreaming TVET. A third disruption occurred that was favourable to TVET teaching especially for IR4.0 was the onset of the Covid pandemic two years ago (in 2019) that compelled the rapid digitization of instruction throughout the education system. These positive developments notwithstanding, some things have changed for the better, but some issues remain. Before we deal with these issues, let us track the unfolding narrative of Malaysia's TVET through these major developments.

The Eleventh Malaysia Plan: Proud of the country's relatively recent economic performance but also facing challenges, internal (such as low productivity, declining export competitiveness, shortage of skilled labour and limited fiscal space) as well as external (such as low commodity prices, falling value of the ringgit, and China's growth deceleration), the 11MP has as its theme people-centred growth. Among its six strategic thrusts is "accelerating human capital development for an advanced nation" with its "game changer" of strengthening governance through unifying accreditation and rating systems, "enabling industry-led technical and vocational education and training" and introducing specializations (Amir, 2017: 3). Efforts would also be made to rebrand TVET. The targets for

Table 11.1 Criticisms of Malaysia's TVET as at 2013

Issue/Challenge	Issue/Challenge Explained
Technical teaching staff shortage	Most of the staff assessed based on their academic qualifications with little importance given to their skills. Those with work experience unwilling to become teachers due to the unattractive salary scheme.
Flexible access to TVET throughout life.	Most who could benefit from it do not usually have access given high operational cost. Less costly alternative methods of delivery of instruction needed, e.g., flexible teaching and learning materials delivered online.
Enculturation of life-long learning.	Needed legislation, institutional structures and redesigned curricula to cater to all members of society to enter and re-enter the world of work. Continuous TVET opportunities for personal and social development to further study. A mindset change needed for a life-long learning culture.
Weak monitoring & evaluation	TVET programs largely supply-driven and lack matching training to available jobs. Training institutions also seldom track the employment destination of their graduates. Outcome evaluation and tracer studies are still lacking. Response from the private sector to industrial training is lukewarm, with private (vocational) training institutions struggling to attract financial support and students.
Concentrated in lower level skills	TVET provision largely concentrated on lower-level skills qualification, with over 70 percent of graduates in Skills Certificates, Levels 1, 2 and 3. Training focus should be concentrated on higher-level. Funding structure does not fully support quality and performance of TVET providers.
Non-homogeneous participation of ethnic groups	Indian youths make up less than 3 percent of the total intake to TVET places offered in the country. The overall participation is dominated by Malays.
Attrition and completion issue	Attrition rates and completion rates of students are of vital concern. Although attrition numbers are not significantly large, some students fail to graduate in time. Need to improve completion rates by taking into account quality and the supervisory system.

Source: Adapted from Ismail and Hassan (2013), Table 3.

this strategic thrust are for labour productivity to grow at 3.7 percent annually, for compensation to employees to reach 40 percent of GDP by 2020, for the monthly median wage to reach RM2,500 by that year. With respect to TVET, the Plan aimed to have 60 percent of the 1.5 million jobs created to be filled by TVET graduates, with 225,000 secondary school leavers enrolling in TVET programmes. And more employees benefit from lifelong learning through the Human Resources Development Fund (HRDF). These initiatives should benefit more than 1,000 TVET institutions from federal ministries, state skills centres, and private providers.

The Pandemic: From January 2021, when Malaysia recorded its first case of infection by the SARS-CoV 2 (COVID-19) virus, it has been in the grip of the pandemic which saw repeated lockdowns (movement controls) and control regulations of varying intensities till today (Pharmaceutical Services Programme, 2020). These lockdowns, in their most intense, saw the "non-essential" workforce working from home or retrenched, small businesses shuttered, and schools and tertiary institutions closed to students. Ernst and Young (2020) expect the virus' impact to be pervasive, directly on some sectors but also on stock markets and consumer spending.

In higher education, the most obvious impact was to migrate all teaching online via e-learning with new platforms. The move to online learning poses both opportunities and challenges. One positive of the move to e-learning is that it mandates familiarity with digital literacy, also a requirement for IR4.0 learning (MIDA, 2021). Thus, MIDA (2021) boasted that "in response to the new normal resulting from the pandemic, Malaysia's higher education institutions have jump-started their initiatives in adopting e-learning approaches (and) Open and Distance Learning (ODL) practices." It also noted the use of MOOCs by several universities. Another possible benefit is to force the higher education sector to adapt to a new regime of working and learning from home. In doing so, a learning management system needs to be put in place, but the arrival of the pandemic has given little time for higher educations to master these.

But online learning has its own limitations. First, not all learning is amenable to online delivery. Laboratory work and some research, especially collaborative, cannot be conducted online, while internships are also difficult by this means. Second, even if teaching can be used, its quality would depend on the length of time online learning has been implemented, the comfort of teaching staff in using this method of delivery, and unfortunately the limited time they had to transfer material meant for face-to-face instruction to content online (Norzaini and Doria, 2021: 80). It is also impossible to monitor students' understanding of and/or reaction to the content of the teaching. This and other reasons have left students and their parents unconvinced that online learning is an effective alternative to face-to-face instruction. Arguably the greatest obstacle to effective learning is the digital divide, with remote rural communities and poor households having no access to internet and hence to online learning.[3] Finally, given the speed with which the virus arrived, most institutions are adjusting by trial-and-error, with implementation record at best mixed (Norzaini and Doria, 2021: 89).

TVET Today

Years 2019 and 2020 saw the end of 11MP followed almost immediately by the arrival of the pandemic to Malaysia. What is the state of Malaysian education, specifically the role of TVET, when 11MP ended? Through the 11MP, TVET was recognized explicitly as vital by the government. For this reason, as of June 2017, the number of TVET institutes was over a thousand, 45 percent of which were public sector enrolling 200,000 students (Amir, 2017). But programmes were delivered by a multitude of ministries and agencies, as shown below (UNESCO-EVOC, 2019).

The formal TVET system of the Ministry of Education begins at the lower education level, basic vocational education (ISCED 2) is offered in public schools in the formal TVET system. On completion after three years, students can proceed to skill training schools or enter the job market. Students can also join TVET programmes after they pass the lower secondary examinations. Taught at Technical Schools and other institutions, students who complete the programme (ISCED 3) sit for the SPM (Sijil Pelajaran Malaysia – Malaysian School Certificate) and go on to pre-university studies or enter the labour market. Students who pass SPM and remain in the vocational stream can also opt to join non-tertiary certificate or diploma programmes (ISCED 4). TVET students who intend to pursue tertiary education (ISCED 5–8) to obtain a Bachelor degree or diploma or advanced diploma can proceed to institutions of the Malaysia Technical Universities Network (MTUN) (Figure 11.1).

· But a plethora of other ministries also offer non-formal and informal TVET programmes. The Ministry of Human Resources offer apprenticeships that combine hands-on industrial training (70–80 percent) with classroom instruction (20–30 percent) at Industrial Training Institutes. The Ministry of Rural Development's Majlis Amanah Rakyat also runs skills training institutes. Vocational training for military veterans is provided by the Ministry of Defence. The affiliated agencies of the Ministry of Works also offer TVET training. Other ministries which offer TVET training are the Ministry of Youth and Sport, Ministry of Agriculture and the Ministry of Rural Development.

In addition to the formal public sector TVET system, private sector TVET institutions that account of more than half the number of total TVET institutions offer programmes with various specializations at various levels. How did such an elaborate system develop and what are its challenges in its current role of preparing the workforce and public for the arrival of IR4.0? In its review of the 11MP, the 12MP recognized that notwithstanding its extensive coverage, the

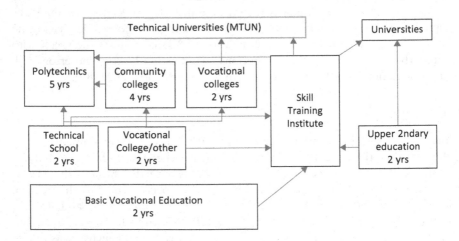

Figure 11.1 Malaysia's TVET Structure.

TVET subsector faced major challenges that saw projected enrolments not met, and TVET students numbering less than half that of those enrolled in academic programmes.[4] Four were cited.

First was uncoordinated governance, particularly significant given the number of agencies offering TVET. Manifestations of this existed in the form of unclear TVET articulation, and only ad hoc industry input. Second was fragmented delivery. Despite a standardized national occupation code, multiple providers each used separate modes of delivery. Public TVETs also offer no specializations while no standardized performance assessment existed. Third, a TVET-trained technologist was not recognized as a professional, with adverse consequences for his/her wages, in addition to having no opportunity to continue to pursue higher education. Finally, competency deficiencies existed among some TVET instructors. Some lacked the technical skills while others lacked industrial exposure.

To remedy these challenges, the Plan intended to strengthen governance, improve the quality of programs and rebrand the entire subsector to increase its attractiveness. In terms of governance, a single accreditation system was proposed as was a single rating system for both public and private TVET institutions. In program delivery, it would be industry-led in partnership with TVET institutions and government. Specializations would be allowed. The number enrolled in TVET would be raised from 163,000 in 2013 to 225,000 in 2029. And TVET students would face clear career choices. Though no explicit mention had been made of IR4.0, the objectives of moving to high value-added production and industry-led training were consonant with it.

What are the outcomes of these initiatives at the end of the Plan? The 12MP concluded, albeit optimistically (12MP: 10–3):

> During the Eleventh Plan, Malaysia continued to record full employment, improved labour productivity and higher graduate employability. Enrolments in schools, higher education institutions (HEIs) and TVET institutions also increased.

However, in the same breath, "the unemployment rate increased and labour productivity declined in 2020", blamed on movement control to deal with the surging pandemic. In addition, only three of the nine targets to empower human capital were achieved. Among these, TVET experienced mixed success. While its graduates enjoyed greater employability, the intake of those who completed secondary school fell short of the projected 225,000 students, "partly due to the public perception of TVET as a second choice[5] and the fragmented TVET landscape" (12MP: 10–3).

With specific reference to preparation for IR4.0, arguably one of the most striking features is the multitude of agencies, briefly described earlier, with responsibility to deliver TVET programmes, making for "overlap and unclear division of responsibilities" (UNESCO-UNEVOC, 2019: 12). And in terms of IR4.0 upskilling, despite detailed job specifications for each level of TVET qualification (UNEVOC, 2019: 15–16), the extent to which each agency is committed among its many priorities is hard to tell, but even with all agencies pitching in enthusiastically, coordinating their content if not pedagogy would pose extremely challenging. Also given the detailed job descriptions for each level of TVET qualification, major restructuring of studies to accommodate IR4.0 subjects would be needed. Steps to remedy these issues were taken, however, when then Minister of Education Mazlee Malik established the National TVET Coordinating Body to bring TVET under one roof (Amir, 2020), a practice since continued in the 12MP.

Further, that "Malaysian teacher education is still dominated by degree courses" (UNEVOC, 2019: 13) also suggests that in hiring TVET instructors, not enough weight has been accorded industry experience (see for instance Bandura and Grainger, 2019; OECD, 2018). And to the extent in-service training is provided, it is "mainly to upgrade the teachers' rank and salary category" (UNEVOC, 2019: 14). Further, the unification of qualifications for both academic and TVET subsectors under 11MP has only served to entrench the dominance of academic qualifications in hiring and promotions (UNEVOC, 2019: 15). The saving grace is the existence of hands-on industrial training combined with classroom instruction, which carries weight for IR4.0 learning.

Confirming the persistence of major challenges, Boo (2018) acknowledged that first, until the launch of 11MP and its successor 12MP, TVET did not receive as much attention as academic education even in government circles. Even after the government endorsed explicitly the important role of TVET, the general perception that those students who enrolled in TVET programmes were those who could not cut it in academic studies continued to prevail. This explained a second challenge – despite the government's grand plans and ample budget, the target enrolment for TVET could not be achieved. As evidence of this shortfall, Boo (2018) estimated that only 7 percent of secondary school leavers enrolled in TVET programmes while only 70 percent of places had been filled. The former number could be compared with Germany's and Switzerland's 70 percent and Singapore's 75 percent. Similarly, dropouts from primary but more from secondary school had fallen but were still significant. Thirdly, he noted that the proportion of high-skill workers had fallen over the past 15 years but the proportion of unskilled workers had increased. Fourth, despite Malaysia enjoying near full employment, youth unemployment was much higher than the national average, with graduate unemployment a major share of youth unemployment.

Many of these issues represent a reprise of those raised in the World Bank SABER Report and by Ismail and Hassan (2013) issued five years ago, showing the absence of major improvements in key areas despite the government's strong support. First is the persistence of a negative attitude towards TVET, together with the continued reliance on academic qualification in hiring instructors and the persistence of graduate unemployment despite an overall situation of full employment. Nevertheless, some improvements had been made. These included efforts to "rebrand" TVET, stronger industry participation of and driven programmes, unification of accreditation and rating systems and coordination between the many ministries and agencies offering TVET programmes. With respect to IR4.0, unfortunately, the extent to which the substance and quality of programme delivery effectively address important priorities is not readily known.

But small sample and/or selective micro-studies can throw some light on these issues. One group of these deals with lecturers' satisfaction with delivering programs, including IR4.0. Not surprisingly these subjective assessments were positive. Thus, in a survey of 100 lecturers in a TVET university, Rahman (2017) found them with favourable views of TVET teaching and learning. Zulnaidi and Majid (2020) found that lecturers' understanding and readiness to teach IR4.0 subjects were high. Similarly, Zulnaidi, Heleni, and Muhammad (2021) found TVET lecturers' understanding and readiness to teach IR4.0 subjects to be high. Others (Ibrahim, Baharuddin and Baharom, 2018: Shafei, Haris and Hamzah, 2018), however, found readiness to be only moderate.

These mixed results notwithstanding, they said nothing about delivery efficacy. For this, we need to turn to surveys of beneficiaries of such teaching – the students. Chalil (2019) reported on a survey by IDC and INTI International University and colleges that 63 percent of students and graduates and 54 percent of parents could not articulate IR4.0 concept. Thirty percent of students were "unprepared for the IR4.0 workplace and 28 percent of students said their only exposure to IR4.0 was in university. Reporting on the same survey, Sani (2020) reported that "students had been too reliant on academic programmes to make them job-ready... not many students took the initiative to take courses to get additional certificates relevant to the IR4.0 workplace." However, another survey of 400 public university students came to the opposite conclusion (Idris, 2019). It found that most had heard of IR4.0 (90 percent) or had some understanding of IR4.0 (70 percent) although up to half felt that their universities could have done better (55 percent) or provided better infrastructure (60 percent).

Implications for Malaysia

In this chapter, two questions were posed the answers to which prompted a third question which has implications for policy. The first question is what is the state of preparedness of Malaysian for the arrival of IR4.0 which the government is active planning for. This question was answered earlier in this chapter, which was, with industry wedded to the low-skill model that saw only 25 percent of the workforce having a bachelor's degree (Noorziah et al., 2015: 226) and employers[6] resisting an increase of minimum wages, partly owing to the pandemic, that Malaysia was not yet in the IR4.0 era. Apparently, Malaysian industry prioritized short-term profits over long-term competitiveness.

The second question, as to whether Malaysia's TVET is at or close to that of leading countries Singapore, Germany and South Korea. The answer is again not quite. Although the Malaysian TVET has seen major improvements with the government pouring resources into this sector, major impediments remain to be overcome. Apart from structural issues of coordination and standardization, arguably the greatest obstacle yet to be overcome is the public perception that TVET is an inferior alternative to academic education. This perception, no matter how much rebranding occurs, so long as industry does not give due recognition to TVET graduates, will persist.[7] Additionally, the content of Education 4.0 is now well-known. But how much has been incorporated into Malaysia's TVET is not precisely known. Nor, despite the positive feedback, is it known how much teaching staff have bought into the radically changed teaching and mentoring methods envisaged.

The answers to the aforementioned two questions prompt a third. Which is, given industry lagging behind the IR4.0 environment, is there any point in TVET upgrading to teach IR4.0 subjects, i.e., to embrace Education 4.0? The answer, to our minds, is still yes. As a learning method it presents a novel way of learning putting much greater focus on industry-learning linkages. It may transform the entire education sector once, through its focus on soft skills, its salutary effects on graduate employability are recognized. It also strengthens familiarity with emerging or skills already in use in technologically advanced countries. And once these skills prove their worth, industry may come around to recognize the desirability of embracing IR4.0.

Finally, reforming Malaysia TVET is certainly in the realm of the possible. Singapore, Germany and South Korea, much praised for their TVET, all offer useful lessons from which Malaysia can learn. Germany recognizes three dimensions of competence: professional competence, that of the individual in relation to his profession, social competence, that of the individual in relation to his social environment and self-competence, which relates to the individual in relation to himself and his own qualities. German apprentices are exposed to a real industrial work situation through TVET very early from the age of 15 in upper secondary education. Apprentices work 70 percent in the real economy with paid wages and 30 percent in formal education. Almost all German apprentices are trained IR4,0. The Relevance of Entrepreneurship Education for Development 15 education and training would be released from their original industry after completing their training (Salman, et al., 2020).

As early as the 1970s, policy formulation in Singapore benefitted from stakeholder engagement through the Tripartite Alliance, made up of the unions representing the workforce as the beneficiaries of TVET, employers as consumers of TVET output, and the government (Cheong and Lee, 2016). This collaboration has produced new institutions to guide TVET in a system which is demand driven, vital to the system's ability to adapt in times of rapid economic change. System wise, TVET started for students in lower secondary education, and TVET students could eventually achieve higher education in an education system that advocates the learning process as a lifelong affair.

In Korea, TVET qualifications are considered equivalent to academic qualifications. Its marketability heavily dependent on labour market requirements, has always matched that of other graduates, given South Korea's focus on modern heavy industry; South Korea being number one in terms of the highest industrial robotic density since 2010, with 631 robotic arms per 10,000 employees, reflecting its significant role in training for IR4.0. A distinctive feature of the Korean TVET system is the high proportion of students in tertiary education.[8] (UNESCO-UNEVOC Korea) As can be seen, there is something to be learned from each country model.

Conclusion

In considering the role of TVET in upskilling the Malaysian workforce in readiness for IR4.0, two questions must be asked. First, what situation is Malaysian industries in with respect to IR4.0 which industrially advanced countries are now implementing? And second, how capable is Malaysia's TVET in fulfilling its intended mission of equipping the workforce with IR4.0 work skills. The answer to the first question is that Malaysia's low-cost, low-wage model has not seen industries operate at the IR4.0 level. The answer to the second is that Malaysia's TVET instruction has likewise not fully transferred IR4.0 skills. These answers lead to a third question which is should TVET wait for industry to catch up before launching into upskilling programmes? Hardly. For three reasons. First, Education 4.0 to cater to IR4.0 represents a new paradigm of learning that is likely to transform education, both academic and vocational, and bring about closer collaboration between learning and the workplace. This alone creates value for society. Second, increasing familiarity with IR4.0 skills among the workforce may precipitate the industry's upgrading to an IR4.0 environment. Even if employers do not see the value of IR4.0, their employees will see it as their future. Third, with the rest of the world moving to IR4.0, possessing the requisite skills enables rapid catch-up with right policy incentives in place, and, at a minimum gives those in the workforce thus equipped to seek employment in countries with IR4.0.

Notes

1 Wawasan 2020 was a vision advanced by Prime Minister Mahathir Mohammad during the tabling of the Sixth Malaysia Plan in 1991. The vision called for Malaysia to advance to the status of a self sufficient industrial nation by year 2020 (Mahathir, 2008),
2 The same article reported that only about 90 of the approximately 500,000 Malaysian SMES were eligible for funding under the national blueprint.
3 Norzaini and Doria (2021: 78) noted that "Although the Malaysian Communication and Multimedia Commission (MCMC) in 2019 reported that the national mobile broadband penetration rate was exceedingly high (120 percent per 100 people), the fixed broadband penetration rate which provides faster and more reliable connectivity was significantly low (approximately 8 percent per 100 people)
4 According to MIDA (2011), as of 30 September 2019, there were 1,325,699 students pursuing tertiary education in public and private higher education institutions. Even excluding the approximately 130,000 international students, the number of local tertiary education students was nearly five times the number of students enrolled in TVET institutions.
5 This problem is not unique to Malaysia. Tamrat (2019) found the same perception in Ethiopia.
6 Malaysian employers have been labelled "stingiest in Southeast Asia on account of wages accounting for only 25 percent of the country's GDP as opposed to 40 percent for Singapore, 84 percent for Indonesia and 76 percent for the Philippines (Renushara, 2022). But this reflects as much the low-wage model Malaysian industry adopted as employers' stinginess.
7 No matter what "experts" advise, (Selangor Journal, 2022)this perception will remain unchanged unless and until evidence exist that TVET graduates are treated as professionals by industry.
8 This is reported to be 22.8 percent compared to 17.5 percent for secondary education (UNESCO-UNEVOC Korea, 2018: 3).

References

Amir, O. (2012) Skills Development in the 21st Century: Malaysian Experience, Downloaded on November 19 from file:///C:/Users/RRAJAH/Downloads/Skills%20development%20in%2021%20Century%20Malaysia%20-Amir%20Bin%20Omar%20MoHR.pdf

Amir, J. (2020) "TVET shouldn't be second choice", *EMIR Research*, February 24. At https://www.emirresearch/tvet-shoulnt-be-second-choice/.

Bandura, R. and Grainger, P. (2019) "Rethinking pathways to employment: technical and vocational training for the digital age", *Conference on the Future of Work and Education for the Digital Age*, Japan 2019.

Bernama (2019) Decline in STEM students worrying, says Mazlee, Bernama, March 18, downloaded on November 29 from http://www.dailyexpress.com.my/news/132551/decline-in-stem-students-worrying-says-maszlee/.

Boo, C.H. (2018) "From TVET in industry 4.0 to reshaping our perception of education", *Malaysiakini*, July 23. At https://www.malaysiakini.com/letters/435396.

Chalil, M. (2019) "Study: More than 63% of students in Malaysia can't explain what IR.4.0 is", *Malay Mail* August 20. At https://www.malaymail.com/news/life/2019/08/20/study-more-than-63pc-of-studens-in-malaysia-cant-explain-what-ir4.0-is/1782417.

Cheong, K.C. and Lee, K.H. (2016) "Malaysia's education crisis: can TVET help?", *Malaysian Journal of Economic Studies*, 53(1): 115–134.

Cheong, K.C., Lee, H.A., Abdillah, N. and Kuppusamy, S. (2013). Malaysia: Workforce Development, SABER Country Report 2013, Washington, DC: World Bank.

EPU (2021a) *Malaysia Digital Economy Blueprint*, Purajaya: Economic Planning Unit.

EPU (2021b) Malaysia (2018). *IndustryWRD: National Policy on IR 4.0*, Putrajaya: Ministry of International Trade and Industry.

Ernst and Young (2020). Economic Impact of COVID19: The Malaysian Experience, downloaded on November 29 from https://myfuturejobs.gov.my/wp-content/uploads/2020/04/COVID-19-Economic-Impact_Malaysia_080420.pdf.

Gerstein, J. (2014). "Moving from Education 1.0 Through Education 2.0 Towards Education 3.0. Experiences in Self-Determined Learning". In M. Blaschke, C. Kenyon and S. Hase (eds.) *Experiences in Self-Determined* Downloaded on November 29 from http://scholarworks.boisestate.edu/edtech_facpubs/10.

Gomez, E.T., Cheong, K.C. and Wong, C.Y. (2021) "Regime changes, state-business ties and remaining in the middle-income trap: The case of Malaysia", *Journal of Contemporary Asia*, 51(5): 782–802.

Ibrahim, R., Baharuddin, S. and Baharom, H. (2018) "Memperkasa TVET: Readiness level of polytechnic lecturers in state transformation 21st century along with Industrial Revolution 4.0", *Paper Presented ate the 8th National Conference in Education – Technical & Vocational Education and Training (CiE-TVET)*: 261–270.Korea

Idris, R, (2019) "IR4.0: An overview of readiness and potential economic effects in Malaysia from millennials' perspective", *World Scientific News*, 118: 273–280.

Ismail A. and Norhasni, A.B. (2014). Issues and Challenges of Technical and Vocational Education and Training in Malaysia Towards Human Capital Development, *Middle-East Journal of Scientific Research*, DOI: 10.5829. idosi.mejsr.2014.19.icmrp.r

Kaur, B. (2019) "Malaysia falls behind in IR4.0 migration", *New Straits Times*, August 29. At https://www.nst.com.my/news/nation/2019/08/514202/malaysia-falls-behind-ir40-migration.

Kozinski, S. (2017) "How generation Z is shaping the change in education", *Forbes*, July 24. At https://www.forbes.com/ sites/sievakozinsky/2017/07/24/how-generation-z-is shaping-the-change?? -in-education/#304059746520.

Lee, R, (2020) "Cover story: IR4.0 in Malaysia: The challenges", *Edge Malaysia*, March 02. At https://www.theedgemarkets.com/article/cover-story-ir4.0-malaysia-challengers.

Lee, W.Y., Tan, S.T. and Sorooshian, S. (2019) "Impacts of industry 4.0 on Malaysian manufacturing(TVETg industries", *WSEAS Transactions in Business and Economics*, 16: 355–359.

Mahathir, M. (1991) "The way forward (vision 2020)", Working paper, tabling of Sixth Malaysia Plan.

Malaysia (2018). *IndustryWRD: National Policy on IR 4.0*, Putrajaya: Ministry of International Trade and Industry.

Malaysian Industrial Development Authority (MIDA) (2021) *Evolution of E-Learning in the Malaysian Higher Education Institutions*, February 8. At https://www.mida.com.my/mida-news/evolution-of-e-learning-in-the-malaysian-higher-education-institutions/.

Mibrand (2017) Ir4.9: *How it's Going to Change Malaysian Manufacturing*, July 28. At https://mibrand.my/ir4-0-how-its-going-to-change-malaysian-manufacturing/.

MIDA (2011). *Annual Report*, Kuala Lumpur: Government Printers.

Miranda, J., Navarrete, C., Nnoguez, J., Espinosa, J.M.M., Montoya, M.S.R., Tuch, S.A.N., Bello, M.R.B., Fernandez, and Molina, A. (2021) "The core components of education 4.0 in higher education: Three case studies in engineering education", *Computers and Electrical Engineering*, 93: 1–13.

Nasir, N. (2019) "Highlights of IR4.0 in Malaysia", EMAG, November 20. At https://www.emag/livehighlights-of-the-ir-4-0-in-malaysia/.

Noorziah, M.S., Abdul Kadir, B.H.R. and Dg. Kamisah, A.B. (2015) "Human resource management roles and skills shortages in Malaysian organisations", *Open Journal of Social Sciences*, 3: 219–226.

Norzaini, A. and Doria A. (2021) "A critical analysis of Malaysian higher education institutions' response towards Covid-19: Sustaining academic programme delivery", *Journal of Sustainability Science and Management*, 16(1): 70–96.

OECD (2018) "Seven questions about apprenticeships: Answers from international experience", OECD Reviews of Vocational Education and Training, OECD Publishing, Paris.

Pharmaceutical Services Programme, Ministry of Health Malaysia (2020). *COVID-19 Pandemic in Malaysia: The Journey*. Petaling Jaya: *A Report*, Ministry of Health Malaysia.

Rahman, N.A.A. (2017) "Teaching and learning of TVET in Malaysia: Insights from academic staff", *Journal of Applied Environmental and Biological Scien*ces, 7(2): 155–157, 2017.

Rasiah, R. (1995) *Foreign Capital and Industrialization*. Basingstoke: Macmillan.

Rasiah, R. (2011) "Is Malaysia facing negative deindustrialisation?", *Pacific Affairs*, 84(4): 714–735.

Rasiah, R. (2020) Industrial Policy and Industrialization in Southeast. In Arkebe, O., Cramer, C., Chang, H.J. and Kozul-Wright, R. (eds) *Oxford Handbook of Industrial Policy*, Oxford: Oxford University Press.

Renushara (2022) "Report: Malaysian employers stingiest in Southeast Asia when it comes to paying employees", *Word of Buzz*, February 18. At https://worldofbuzz.com/malaysian-employee-dubbed-the-stingiest-in-southeast-asia-in-terms-of-employees-wages/.

Salman, M., Shabbir, M.S., Arshad, M.N., Mahmood, A. and Ali Sulaiman, M.A. (2020) 4th Industrial Revolution and TVET: The Relevance of Entrepreneurship Education for Development, *Opcion*, 35(24): 11–21.

Sani, R.M. (2020).Mission ahead: Future-proofing education, downloaded on November 29 from https://www.stem.org.my/my/index.php/en/resources/stem-articles/news-malaysia/mission-ahead-future-proofing-education.

Selangor Journal (2022) *Time to Get Rid of TVET's Second Class Education Label – Experts*, February 21. At https://selangorjournal.my/22/02/time-to-get-rid-of-tvets-second-class-education-label-experts/.

Shafei, S., Haris, M.H.H., and Hamzah, Z. (2018) "The readiness of POLIMAS lecturers in the challenges of industrial revolution 4.0", *Paper presented at the 8th National Conference in Education – Technical & Vocational Education and Training (CiE-TVET)*: 577– 582.

Sony, M. (2020) "Pros and cons of implementing Industry 4.0 for the organizations: A revigorjournal/ew and synthesis of evidence", *Production & Manufacturing Research*, 8: 1, 244–272, DOI: 10.1080/21693277.2020.1781705.s-.

Tamrat, W. (2019) "Universities vs TVET – are attitudes the problem?" *University World News*, March 20. At https://www.universityworld news.com/post.php?story=20190315095852544.

Techwire Asia (2019) Why Malaysian SMEs are struggling with IR4.0, Techwire Asia, August 21, dowloaded on November 2022 from https://techwireasia.com/2019/08/why-malaysian-smes-are-struggling-with-industry-4-0/.

Teok, S. (2022) "Malaysian firms resist minimum wage hike despite spending slump", *Straits Times*, February 8. At https://www.straitstimes/asia/se-asia/malaysian-firms-resist-minimun-wage-hike-despite-spending-slump.

TMOne (2019) *Industry 4.0 Is Here. Where Are the Manufacturers?*, April 10. At https://www.tmone.com.my/resources/thinktank/article/industry-4-0-is-here-where-are-manufacturers/.

UNESCO-UNEVOC International Centre for Technical and Vocational Education and Training (2019) "TVET Country Profile Malaysia", *Bonn*, June.

UNESCO-UNEVOC Korea Research Institute for Vocational Education (2018) "TVET country profile Korea", *Bonn*, November.

Zulnaidi, H.and Majid, M.Z.A (2020). Readiness and understanding of technical vocational education and training (TVET) lecturers in the integration of industrial revolution 4.0. *International Journal of Innovation, Creativity and Change*, 10(10): 31–43.

Zulnaidi, H., Heleni, S. and Muhammad, S. (2021) Effects of SSCS Teaching Model on Students' Mathematical Problem-Solving Ability and Self-Efficacy, *International Journal of Instruction*, 14(1): 475–488.

12 The Role of STRAND in Malaysia's Industrial Transformation

Naguib Mohd Nor and Nazreen Mohd Nasir

Introduction

Humans are living in a time of countless breakthroughs in areas ranging from autonomous vehicles to gene-editing systems and it is the fusion of these technologies from various domains that differentiate Industry 4.0 from the previous three industrial ages. From an economic standpoint, Industry 4.0 has the potential to raise global income levels and improve the quality of life for everyone globally due to new products and services that these technologies have enabled e.g. Grab/Uber car-hailing service and vertical farming. What this means is that transportation and logistics costs will drop and global supply chains will become more effective, all leading to the diminishing cost of trade which will open new markets and drive economic growth.

The full potential of Industry 4.0 remains to be fully realized and these unlimited technological possibilities are fast emerging as we connect billions of people and devices worldwide. It is through the exchange of diverse ideas which has been exponentially scaled via the Internet that technological advancement has been accelerating. Therefore, a great way to think about Industry 4.0 is that it is a direct effect of increased globalization. In this world, globalization will not disappear but rather deepen. If in the past global integration grew as trade barriers came down, it will now rely on the connectivity of national digital and virtual systems and the related flow of ideas and services.

All this is disrupting the global economy and the manufacturing sector is already seeing the effects where automation and localization are beginning to displace traditional supply chains, which means yet again that going forward, competition will depend ever more on the ability to innovate. STRAND has observed this trend in its work with the Malaysian supply chain.

Economics has always been inextricably linked to disruption. New technologies and production methods create innovative ways of deriving new value and in some cases superseding old ones. This concept was pioneered and disseminated into the mainstream by the Austrian economist Schumpeter (1942) who termed it as "Creative Destruction" and no other manifestation of the idea is demonstrated as strikingly as right now with the advent of Industrial Revolution 4.0 (IR4.0) technologies. Nevertheless, it needs to be highlighted equally, if not more,

DOI: 10.4324/9781003367093-12

the damaging dimension to "Creative Destruction" as its beneficial counterpart because although disruption has the potential to create new employment opportunities and education for the good, there is also the danger of rising inequality through a "Winner-Takes-All" global economic scenario that pushes less developed economies and low-skilled human capital out of competitive positions. Examples include the Elon Musk Effect with the mass introduction of autonomous electrical vehicles (Fox, 2020) Netflix and Spotify soaking up the market share previously held by incumbent players once thought too big to fail (DW, 2018; Haridy, 2019) and how 3D printing is reducing the need for materials and parts that were normally sourced globally (Manufacturing Today, 2020), to name a few. Such scenarios are now being accelerated by the COVID-19 pandemic (Plummer, 2020; McKinsey & Company, 2021; Tan et al., 2020). Malaysia's aerospace supply chain is already beginning to grapple with these challenges.

Companies and even whole industries have always found it difficult to keep up with disruption with the overriding notion that it had more time, such as the case with Kodak, Blackberry, Blockbuster and many other established brands considered too big to fail until it did. Why is it difficult for companies to get in front of disruption? Ray Kurzweil from the Singularity University, came up with an explanation for which he coined the term "Law of Accelerating Returns" where he says that evolutionary systems, like information technology, produce exponential changes. This happens because one generation of technology builds on and accelerates the returns of previous generations e.g. the development of the internet has led to the wireless broadband, which then fast-tracked mobile apps development and has led up to cloud computing in its current form. These accelerating returns produce exponential curves in a system's fundamental measures which means the measures of power and speed tend to double at consistent intervals with huge reduction in costs i.e. Moore's Law.

This is why it is difficult to stay ahead of the curve with disruption because companies, especially those that are asset-heavy, tend to think linearly due to the fact that so many considerations need to be accounted for, limited resources being one of them. As a result, these companies are slow but agile disrupters that have no such baggage are building the next industry-creating experiences using emerging technologies and riding the next wave of exponential change thereby pulling ahead from the rest of the competition. STRAND has been developing solutions to allow for technology asset leasing to become an option for companies to lighten their capital load.

From all of this developmental progress comes with it a pattern of increased intensity of energy usage. As societies become more prosperous, it is followed with a continuous quest for higher energy use which is not necessarily being rationally utilized. According to Vaclav Smil, mankind has experienced three major energy transitions and is only now trying to kickstart a fourth. The first phase saw the mastery of fire, which has allowed the liberation of energy from the sun by burning wood i.e. plants and to cook the meat of the hunted animals. Second phase was farming, which converted and concentrated solar energy into food thereby freeing people for pursuits other than sustenance which was hunting.

It was also during this era where farm animals and larger human populations also supplied energy in the form of muscle power, to till the farmland. The third energy transition was industrialization and with it came the rise of fossil fuels e.g. coal and oil, and energy production became the domain of machines, as such the coal-fired power plants. In an increasingly warming world, this is unsustainable. Mankind must look for more rational ways of consumption and make a transition to a less energy-intensive society, from highly concentrated fossil fuels to more dispersed renewable sources such as biofuels, solar and wind farms. How could the world get in front of this sustainability issue then? As Kate Raworth articulates in her book *Doughnut Economics*, there needs to be a shift from the current economic paradigm which is based on a value chain model with the sole goal of profit maximization and unchecked growth. Besides the crude exploitation of the natural environment and resources, it also leaves a negative impact onto society.

Raworth's "Doughnut of Social and Planetary Boundaries" provides the framework to model the alternative to grow healthy people, a healthy economy and maintain a healthy planet. Inside the Doughnut Hole is the Social Foundation, which comprises 12 basic human needs e.g. water, food, justice and education. Around the rim of the Doughnut are nine planetary boundaries which represent Earth's ecological ceiling. If things like ocean acidification, land conversion, air pollution, and climate change are overshot, that could result in irreversible damages to the planet. Therefore, the ideal space for the economy is to be within the space between the Social Foundation and the Ecological Ceiling which means that there is a balance between satisfying society's needs, as well as maintaining the Earth's health.

In order to make the economy sustainable, it has to be Circular which is the opposite to the current linear "Take, Make and Dispose" consumption model. Achieving this transition requires a fundamental system redesign as the Circular Economy is essentially a simulation of the natural ecosystem process of cyclical resource and towards zero waste. The ecosystem of the Circular Economy works in closed loops of resources which means companies will need to take responsibility for any materials in its products that go beyond sale and use. COVID-19 has accelerated this way of thinking particularly in the aerospace industry where aggressive targets are being set to reduce the carbon footprint of the industry. STRAND is involved in the development of such technologies with new aerospace OEMs. Consequently, this chapter explains the role STRAND has played so far to stimulate the absorption of IR4.0 technologies among Malaysian industries.

Theoretical Considerations: Towards a Sustainable Economic Growth Model

The importance of galvanizing manufacturing in medium and large economies has received serious attention since the First Industrial Revolution began in Britain (Reinert, 2007). Gerschenkron (1062) and Kaldor (1967) subsequently argued over the role manufacturing plays in structural transformation from low to high value-added activities. Rasiah (2020a, 2020b) differentiated the contrasting outcomes of industrialization initiatives among the South East Asian market economies. Key to industrial upgrading is the role of technology and innovation Schumpeter (1934, 1942). However, as countries seek to unlock policies to spearhead industrialization important epochal changes in technological regimes are key to comprehend with the most recent being driven by digitalization and IR4.0 technologies.

Any attempt to comprehend IR4.0 will have to address the United Nations (2016) Sustainable Development Goals (SDGs), which seeks to meet the multiple pillars of sustainable development targeted at planet earths capacity to contain climate change and global warming while offering the opportunity to enjoy improvements in the standards of living and alleviating poverty. Consequently, a sustainable economic growth model is a key requirement for any nation to survive in this post-COVID-19 pandemic world (Rasiah, Gopi Krishnan and Azleen, 2022).

The growth and development of a nation state's economy will not be sustainable by just fulfilling the short-term or medium-term economic goals alone because of the interconnectivity of all the various elements and factors. Promoting only an economic or profit agenda without considering social and environmental impacts would place the growth and development trajectory in the wrong path (UNESCO, 2021). Therefore, economic growth and development has to be based on the Sustainable Economic System Framework which is built on three dimensions: (i) Economic; (ii) Social; and (iii) Environment. Essentially, the framework serves to highlight the interdependency of the economic, social and environmental dimensions in achieving sustainable economic growth. If one or two of the dimensions were developed but the others were neglected, the inter-dimensional link would break and the development will be rendered unsustainable. The distinct dimensions are mutually reinforcing (see Rasiah, 2019).

A sustainable economic growth and development model in a progressive nation state would therefore require an equitable and non-conflicting development in all dimensions. This entails trade-offs between multiple objectives. The different elements in each dimension can be thought either as an economic, social or environmental factor or outcome. Nevertheless, what is considered the outcome can also have a reverse effect of influencing related causal factors, and hence the link between the causal factors and the outcome in the framework.

Connectivity: An Essential Driver of Competitiveness

The age of IR4.0 and the subsequent Globalization 4.0 will see an increase in competition by a several notches. The principle of "Creative Destruction" that was advanced by Schumpeter (1942) will experience a radically disruptive restructuring, destroying capital advantage at a higher rate than ever before it facilitates the absorption of disruptive technologies into global networks. Siloed or truncated approaches to industry development are no longer sustainable as cooptation built on a higher dimension of competition and cooperation becomes the prevalent method for market entry. Connectivity in the context of IR4.0 has therefore become the pre-requisite to confronting competitiveness.

Successful investment models must consider the entire ecosystem. Investors need the support of the government to be able to profile the risks and returns based on an ecosystem level view which gives them a clear understanding of the synergy and interdependency of the ecosystem players. This is best managed therefore via a public-private partnership (PPP) framework involving multiple stakeholders and robust governance, the latter to be managed also via technologies such as big data and internet of things (IoT) to ensure competitive outcomes through effective empowerment, transparency, timeliness, accountability and equitable distribution of rewards.

Table 12.1 Gaps in the Malaysian Manufacturing Supply Chain

Cost Items	Challenges	Example Causes
Design, Process and tool optimization	• Service not optimized • Low value add making quote not competitive	• No embedded engineering capability • No technology development capability • No MES, ERP, PLM capability • Costly equipment
Machine hours	• Quoted hours not optimized • Machine hour price uncompetitive	• Low average years of experience and not to Industry4.0 standards • Uncompetitive financing
Man hours	• Quoted hours not optimized • High man to process/ machine ratio	• Low average years of experience and not to Industry 4.0 standards
Material cost	• Purchase rates not optimized • Inefficient recycling	• Low purchase volumes • Material net shape supply not available
Utilities	• Purchase rates not optimized • Not green	• Individual companies small scale consumption • Small companies lacking know how
Logistics	• End to end costing not available • Inefficient logistics costing	• Lack of Industry4.0 logistics capability • Lack of value added services e.g. packaging • Lack of flexible multimodal services
Treatment	• Certain services not available • Inefficient pricing	• Lack of third-party providers • In house treatment not fully utilized
Consumables	• Purchase rates not optimized • Not green	• Individual companies small scale consumption • Small companies lacking know how
Technology	• Not lean • Technology disruption (e.g. 3D printing)	• No access to world class standards • Lack of R&D capability

Source: MARA Aerospace and Technologies Sdn. Bhd. and Strand Aerospace Malaysia Sdn. Bhd. (2019b).

STRAND is currently engaged in projects that are oriented towards developing data systems and tools which will enable Malaysian companies and organizations to achieve a workable level of connectivity with industries. At the moment, the aerospace industry is being leveraged on to mature the nation's industrial capability to be able to compete in this environment as it demands high levels of global connectivity from its supply chain constituents from the outset. That said, there are still challenges faced by the current Malaysian manufacturing supply chain in serving the aerospace industry. Some of these challenges are listed in Table 12.1 and is understood to be in terms of capability and capacity which will need to be addressed for the supply chain to be commercially competitive (MARA Aerospace and Technologies Sdn. Bhd. & Strand Aerospace Malaysia Sdn. Bhd., 2019a).

Referring to Figure 12.1, the first three stages i.e. Maturity 1 to Maturity 3 are in reference to a more traditional view of an industrial ecosystem. Here, the ecosystem matures through the establishment and growth of anchor companies. For Malaysia, this is often represented by foreign direct investments (FDIs) complemented by a local supply chain of small- and medium-sized companies (SMEs and MEs). Maturity level 4 however attempts to illustrate a view of Industry 4.0 in its most matured form. Here, smaller and more agile businesses provide cutting edge technology and capabilities as part of a flexible network for production. Connectivity represented by the lines connecting the boxes in this scenario therefore defines the industrial ecosystems competitiveness. Connectivity here represents among other things: (i) lines of communication and data transfer; (ii) frameworks of standards and values; (iii) supply chain interfaces; and (iv) global networks of innovation (MARA Aerospace and Technologies Sdn. Bhd. & Strand Aerospace Malaysia Sdn. Bhd., 2019b).

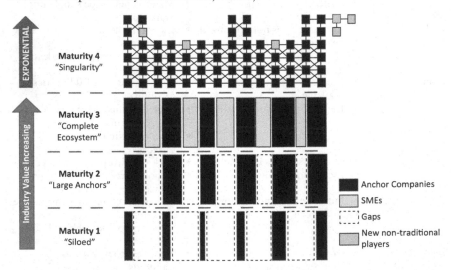

Figure 12.1 Industrial Ecosystem Maturity in High Value Manufacturing Systems.
Source: MARA Aerospace and Technologies Sdn. Bhd. and Strand Aerospace Malaysia Sdn. Bhd. (2019b).-FIG_SRC-

The Role of STRAND in Industrial Transformation

STRAND specializes in the provision of high-value design, analysis, training and turnkey consulting services for aerospace and non-aerospace industries. The company's core engineering capabilities has its roots in the aerospace industry and has served the major OEM such as Airbus and Boeing together with its suppliers on both commercial and military aircraft projects. Leveraging on its aerospace engineering experience and knowledge, STRAND has also applied its services to industries outside of aerospace such as infrastructure, energy, maritime, rail and automotive industries delivering high quality solutions through the application of aerospace rigor, project management and data analysis methods.

Like many of STRAND's engineering services partners in Europe and Asia (e.g. ATKINS, Tata Consulting and ALTRAN), the company began providing non-engineering consulting services off the back of its engineering services and training businesses. This was based on several pull factors:

- The company's global engineering services clients looking for expansion opportunities in emerging markets.
- The company's own need to create more high value aerospace activities in Malaysia to consume its engineering services.
- The local industry seeking support to bridge into high value technology industries.
- The global trend of traditional management consulting companies such as McKenzie and Deloitte developing engineering capabilities in response to similar growing requirements.

STRAND has leveraged its extensive knowledge of aerospace best practices, supply chain mechanisms, OEM requirements and its international network of industry specialists and subject matter experts, to provide turnkey consultancy to the government, government-linked companies (GLC), multinational companies (MNC), large companies (LC) and SMEs since 2006 from which the company has built a repository of industrial insights. The consulting that STRAND provides include:

  i Product and process design.
 ii Business modelling and contract negotiations.
iii Training programs development and delivery.
 iv Technology transfer management.
  v Technical capability development.
 vi Supply chain development.
vii Industrial land and property master planning, development and marketing.

All the above is aligned to the governments push towards high value manufacturing (HVM) in Malaysia and the increasing pressure by the OEMs to reduce manufacturing costs and value whilst keeping up with production schedules.

High value in this sense refers to the value of the manufactured parts being a function of the materials and the value-added processes. HVM industries such as aerospace are more ready for operations to be transformed into Industry 4.0 due to its greater product engineering understanding as well as manufacturing process design ownership. Low-value manufacturing does not have the same level of capability and understanding thus making it difficult for the companies to add value to the product or process by leveraging on Industry 4.0 technologies.

In several of STRAND's consulting projects, clients had properties which since the late 2000s have been challenging when it comes to creating value. This spurred the expansion of STRAND's services to include industrial park master planning. With many of these developments, the key challenge was to narrate a convincing high technology industry proposition that could inform the master plan design. Most master planning services in Malaysia are unfamiliar with such propositions having traditionally been focused on commercial and residential mixed development. This Industrial-Land Development experience that STRAND began accruing allowed the company to be able to understand more intimately the synergies between the different elements of the industrial ecosystems. From here, STRAND developed its ecosystem investment model which it applies to all projects.

The industrial ecosystem investment model is composed of five investment value elements: (i) Technology, (ii) Human Capital, (iii) Industry Ecosystem, (iv) Development and (v) Land, as illustrated in Figure 12.2. These value elements

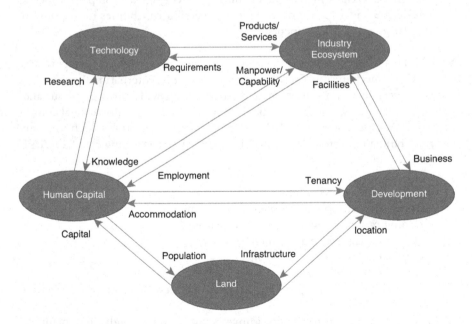

Figure 12.2 Dynamics Between the Different Values Within an Ecosystem.
Source: MARA Aerospace and Technologies Sdn. Bhd. and Strand Aerospace Malaysia Sdn. Bhd. (2019b).-FIG_SRC-

interact with one another to produce ecosystem-level values through the various synergies. All of these together constitute a fertile investment landscape to enable STRAND's clients to leverage on the synergies. The models of the various value elements are built from data and insights derived from the consulting projects that STRAND has undertaken over the years which is currently termed as the Stratified Investment Model (SIM) (MARA Aerospace and Technologies Sdn. Bhd. & Strand Aerospace Malaysia Sdn. Bhd., 2019b). As shown in Figure 12.3, the SIM serves as the investment strategy that captures ecosystem-level value by capitalizing on the synergies with cluster development being at the core of the strategy and entrepreneur enablement as the critical success factor of the cluster.

The solutions for the SIM are being iteratively matured across STRAND projects. STRAND engages companies and institutions over several phases of each of their projects development refining and adding complexity to the SIM. The SIM measures the maturity of the ecosystems by considering both the business models of the individual organizations and its interdependencies. STRAND is currently engaged in government projects which now have also contextualized the SIM as a method for enabling state and national Industry 4.0 strategies. Examples of the SIM value propositions are presented in the following sub-sections.

SIM Industrial Value Proposition: UMW into a Tier 1 Global Aerospace Supplier

UMW Group is a multi-million industrial conglomerate and one of Malaysia's foremost public-listed companies. The company's roots began in 1917 as an auto repair shop in Singapore which then moved on to become a distributorship for automotive parts under General Motors, a household name in the automotive industry in the United States and by 1927, the rapidly expanding business was consolidated into an establishment called United Motor Works Pte Ltd, the predecessor to the UMW Group as it is known today. The company has multiple businesses under its various subsidiaries that are (i) automotive, (ii) equipment, (iii) manufacturing and engineering (M&E) and its most recent (iv) aerospace.

How the aerospace division came to be was through the M&E division which represents the core of UMW's manufacturing capability. At the start, its capabilities were manufacturing and assembly, testing and after sales support but the Group identified that in order for UMW to continue to remain competitive, it needed to have capabilities that cover the early-stage product realization values i.e. design and certification capabilities. This was a challenging proposition considering the contract manufacturing businesses M&E were involved in at the time. The Rolls Royce (RR) Fan Casing contract provided a higher-level business and product vantage point with a world-class partner invested in the transfer of its high-level capabilities to UMW. Thus, UMW M&E was able to increase its value-adding capability through increasing its internal value engineering capabilities and create, via an existing business, a new business unit i.e. the aerospace division (Strand Aerospace Malaysia Sdn. Bhd., 2016).

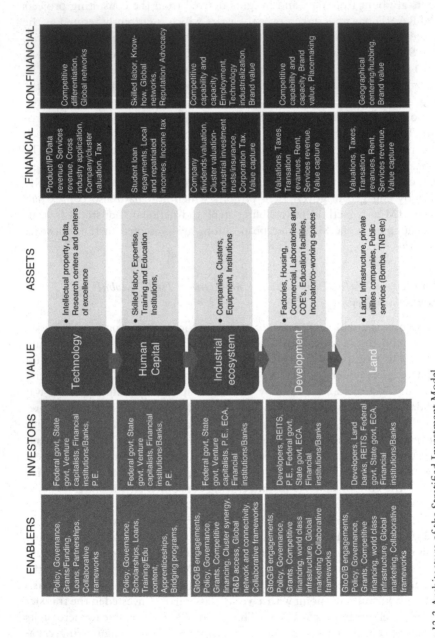

Figure 12.3 Architecture of the Stratified Investment Model.
Source: MARA Aerospace and Technologies Sdn. Bhd. and Strand Aerospace Malaysia Sdn. Bhd. (2019b).-FIG_SRC-

Through STRAND's consultancy services, the transformation journey involved strategic business planning, quotation responses to the OEM, OEM negotiations, factory and industrial park design and tailored human capital programs. In 2015, UMW M&E successfully converted the RR Fan Casing contract and even though it presented a challenging business case due to very high capital investment cost with tight margins, it positioned UMW as the first and only Tier 1 engine manufacturer in the country. This positioning was necessary in order for UMW to:

i   Be able to have access to technology and capability development via RR.
ii Be able to create additional businesses to establish UMW's HVM footprint and transform its manufacturing business and revenues.
iii Be able to spur the growth of the Serendah land ecosystem, which it owns, in order to generate value and revenue.

To sustain this positioning in the coming decades, STRAND recommended that UMW pursue the following:

i   Explore Joint Venture (JV) opportunities with specific and complementary capabilities as well as localization opportunities to be part of the Fan Case supply chain.
ii Establish partnerships, both local and international, to create R&D capability for new technology development and acquisition.
iii Seed the Serendah industry ecosystem with both Domestic Direct Investors (DDI) and Foreign Direct Investors (FDI).

SIM Technology Value Proposition: R&D Platform Management

The establishment of an R&D capability is among the cornerstones needed for any company to be competitive going forward. The world being as highly connected as it is intensifies the flow of information across organizations and borders, highlighting the growing importance of global R&D collaboration frameworks to identify which technologies will yield the highest commercialization value and to bring it to market as quickly as possible.

Figure 12.4 shows an R&D collaboration construct between RR Seletar in Singapore and Malaysian institutions that STRAND was a part of as the Engineering Services provider and Project Manager. This collaboration successfully delivered tools designed and built in Malaysia for use in the RR fan blade manufacturing facility. The project was to modularize a set of heavy tools for a critical tool changeover process taking up to 45mins. The newly designed tools managed to reduce the tool change time by about a third. This construct between RR, the Aerospace Malaysia Innovation Centre (AMIC), the German-Malaysian Institute (GMI) and Nottingham University (NU) ran for approximately eight months and was a first for all involved. AMIC was the project owner and played an overall coordinating role, STRAND was responsible for the overall technical delivery and requirements capture, NU the tool concept design and GMI the prototyping. The planned delivery timelines were exceeded due to the sharp learning curve but STRAND's facilitation was crucial in ensuring that the design data points and technical requirements were communicated properly. Once the design process was completed, the 3D model was handed to GMI for machining and assembly which went through a few loops before the final product to specification was achieved.

STRAND was also the project manager for the development of the Virtual Reality Immersive System Training for Aerospace Manufacturing (VIRISTAM) as shown in Figure 12.5 (University of Malaya, National University of Malaysia, MARA Aerospace and Technologies Sdn. Bhd., Aerospace Malaysia Innovation Centre, 2014). The VIRISTAM programme prototypes the use of virtual reality in the context of training in aerospace manufacturing. The technology and associated training programmes developed were tailor-made for each of the aerospace companies depending on their operational and process needs.

The capabilities associated with conducting such R&D management is as follows:

i   To be able to conduct a needs analysis and identify the pain points of the companies and from there, formulate the project objectives.

ii To be able to put together a strategy for the development with an achievable timeline.

iii To be able to manage the integration of the technology into the company's process line.

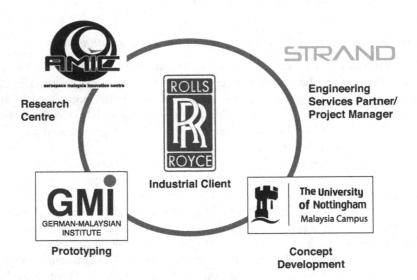

Figure 12.4 R&D Collaboration between Rolls Royce Seletar and Malaysia.

Source: University of Malaya, National University of Malaysia, MARA Aerospace and Technologies Sdn. Bhd., Aerospace Malaysia Innovation Centre (2014).-FIG_SRC-

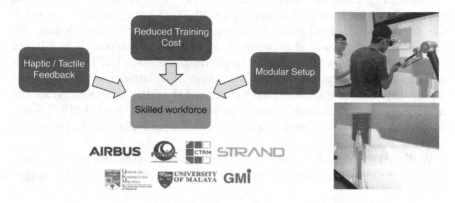

Figure 12.5 The R&D Collaboration Construct for the VIRISTAM Project.
Source: University of Malaya, National University of Malaysia, MARA Aerospace and Technologies
Sdn. Bhd., Aerospace Malaysia Innovation Centre (2014).-FIG_SRC-

SIM Human Capital Value Proposition: Tailored Human Capital Program Development

By entering the aerospace industry as a Tier 1 supplier to RR, UMW needed to establish and manage competency in line with the long-term partnership (25+5 years). UMW would need to be equipped to be able to respond to RR requirements that may go beyond the scope of the initial Fan Casing product (Strand Aerospace Malaysia Sdn. Bhd., 2016). STRAND worked with UMW's top management and RR to define the requirements as such:

i   To adopt a world-class aerospace culture (deriving from RR own "High Performance Culture" initiative)
ii   To have the right level of know-how and tacit knowledge (replicating the level of competence development and management built into the RR apprenticeship programs)
iii To acquire the RR-specified high level of technical capability (as extracted and transferred by STRAND from RR's shop floor in the UK)

STRAND played a central role in designing the Human Capital programme and Learning Management System (LMS) for UMW Aerospace that leverages on its Professional Development Centre (PDC) solution (PDC, 2014). The PDC was setup by STRAND in 2011 as a part of its role in the Economic Transformation Program (ETP). The purpose of the PDC is to bridge the industrial competency gap to increase the employability of local graduates. Three example PDC graduate/technician programs are: (i) Ensuring Employability Program (EEP), (ii) the MARA Learning Institute Program (IPMA) and (iii) the Structured Industrial Intervention Program (SIIP).

The EEP trains and develop talents to be "industry-ready" to perform engineering work upon completion (Figure 12.6). The programme ensures that the talents are recruited through an apprentice-like curriculum focusing on technical knowledge as well as communication and leadership skills to enhance employability. The programme was based on the Airbus UK Higher Education Apprenticeship program which was done in collaboration Glyndwr University (Glyndwr) in the UK. STRAND evolved the syllabus of Glyndwr for the Malaysian industry requirement and then worked with BAESYSTEMS, RR and MBDA to both provide apprenticeship learning experiences and deliver part of the education content.

With the IPMA, STRAND developed short-courses to cater for industry needs utilizing the many MARA institutes as the execution vehicle. STRAND engaged

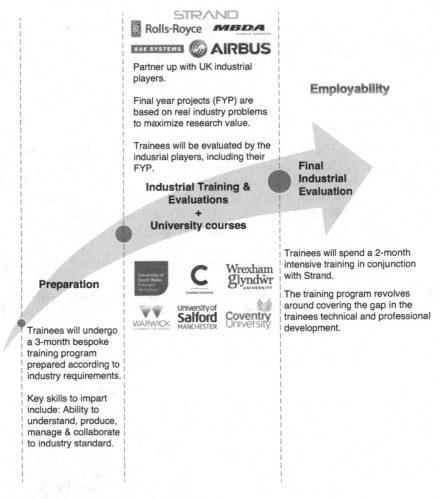

Figure 12.6 Module Structure for the EEP.
Source: PDC (2014).-FIG_SRC-

the companies and functioned as the consultant to assess the requirements and design the tailored industry training plan.

Finally, the SIIP is an intervention programme which provides project management and analysis for undergraduates going through industrial placement (Figure 12.7). The SIIP can be plugged into the programme to be the focal point between the university and the students. The students on this placement programme receive analytics of their development via a Personal Development Plan (PDP) which enables the institution to monitor the students' performance.

STRAND's industry engagement experience is highly strategic and therefore the implementation of the programmes were not built on the BAU (Business-As-Usual) framework but rather transformative e.g. when UMW Aerospace was created, STRAND's specialist teams were brought in to manage almost every aspect of the conversion and launch implementation of the project and then handed it over to the UMW team so that UMW was able to:

i   Be able to grow effective partnerships within this new value chain.
ii Be able to respond to industry in as far as response time.
iii Have a network of subject matter specialists.
iv Have a high degree of flexibility when it comes to implementation.
v   Have industry-level knowledge and capability.

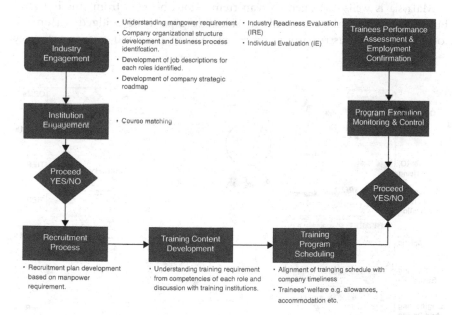

Figure 12.7 Design and Execution of STRAND's Human Capital Development Programme.
Source: PDC (2014).-FIG_SRC-

SIM Industry Value Proposition: Supply Chain Development

The Malaysian aerospace industry supply chain comprises various tiers and support functions as shown in Figure 12.8. Tier 1 companies such as Spirit Aerosystem, UMW Aerospace, CTRM and General Electric create workflow and anchor physical ecosystems. STRAND worked for SME Corporation Malaysia (SMECorp) to deliver the Aerospace SME transformation programme (EPP8) aimed at transitioning non-aerospace SMEs into the aerospace industry through a program called the Toulouse Accelerator Program (TAP) that involved batches of ten selected local SMEs to undergo training delivered by Master Trainers from the Airbus Air Business Academy (ABA) and STRAND comprise webinars, classroom training, workshops, evaluation as well as field trips to Airbus, Airbus supplier facilities, and the Airbus A380 and A350 Final Assembly Line (MARA Aerospace and Technologies Sdn. Bhd. and Strand Aerospace Malaysia Sdn. Bhd., 2017).

The programme handheld the participating SMEs into getting certified and provided the tools to be competitive in the Airbus global supply chain. The Airbus supplier quality management system and process was the core to this program. This system benchmarks SMEs' HR, production, supply chain and technology management processes to Airbus specifications (Figure 12.9). The evaluations carried out during the programme revealed many gaps in the Malaysian SME's capabilities which the programme started addressing (Figure 12.10).

Malaysia is well-positioned to reap tremendous benefits from this initiative but as mentioned, there are still considerable gaps to be bridged within the local aerospace industry in order to establish a complete and mature aerospace

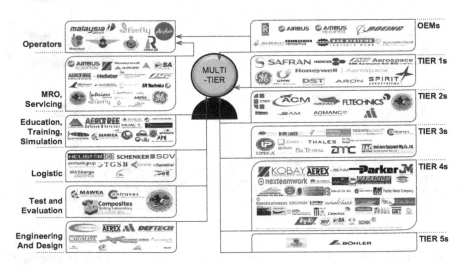

Figure 12.8 The Malaysian Aerospace Supply Chain.
Source: Strand Aerospace Malaysia Sdn. Bhd. (2016).-FIG_SRC-

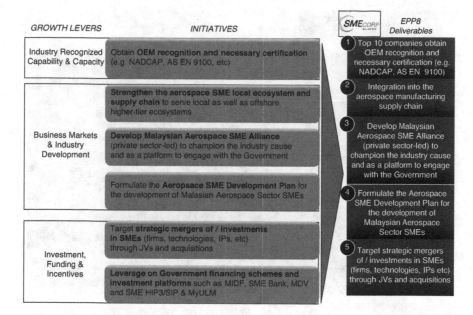

Figure 12.9 EPP8 Aerospace SME Initiative.
Source: Strand Aerospace Malaysia Sdn. Bhd. (2016).-FIG_SRC-

ecosystem. To bridge these gaps, it was recommended that the companies (MARA Aerospace and Technologies Sdn. Bhd., 2017):

i   Adopt a robust Quality Management System (QMS) that is aligned to the OEMs' QMS which cover important areas such as process control, leadership, technology, lean manufacturing, and skills planning and development.

ii   Adopt a system to monitor the regulations set by ICAO, EASA, FAA and DCA that are being enforced or mandated.

iii   Register with the International Aerospace Quality Group (IAQG) and be familiar with the *Supply Chain Management Handbook* (*SCMH*).

iv   Practice Continuous Improvement that is supported by engineering capabilities and commercially driven R&D.

v   Implement an environmental management system (EMS) within their organizations to meet the environmental regulations.

vi   Be aware of the impact on OEMs and their supply chain as a result of increasing air traffic especially in the Asia Pacific region which are driven by the demand for lower cost and access to markets.

SIM Land and Development Value Propositions: Industrial Park Development

UMW contracted STRAND to also develop the master plan for the UMW HVM Park (MARA Aerospace and Technologies Sdn. Bhd. and Strand Aerospace

Figure 12.10 SME Corp delegation in front of the Airbus Leadership Campus.
Source: MARA Aerospace and Technologies Sdn. Bhd. (2017).

Malaysia Sdn. Bhd., 2017). Located in the Serendah district of the state of Selangor, the UMW HVM Park comprises 861 acres of industrial park development. It is a smart city established as a one-stop centre, providing holistic solutions for high tech industries. At the heart of this ecosystem is the UMW Aerospace Fan Case Manufacturing plant which occupies only 20 acres of the overall land.

Malaysia Airports Holding Berhad (MAHB) contracted STRAND for a similar master planning exercise for the aero manufacturing and MRO land parcels in Subang and Sepang Airports. The Northern Corridor Implementation Authority (NCIA) contracted STRAND to develop the master plan of the Sidam Logistics and Aerospace Manufacturing Park (SLAM) adjacent to the proposed Kedah International Airport (KXP). The development of these master plans was based on detailed market studies and suitability of the developments to services specific sections of the aerospace industry. Most of these projects then evolved to involve STRAND in marketing, branding and business plan execution.

STRAND's ecosystem and master planning and UMW Land's execution of the plan resulted in significant increase in the land valuation of Serendah in 2019 (AmInvestment Bank, 2019), which was stagnant for many years prior to that. In context of the SIM, in Serendah, UMW is the anchor workflow creator, facilitating technology and human capital development as exemplified by the Fan Casing factory. Subsequently the government has continued to engage UMW as an industrial developer with joint investment via programs in R&D and human capital development. The UMW HVM Park meanwhile provides an ideal focal point for the regeneration of Serendah for the state of Selangor.

Conclusion

The chapter provided a snapshot of the kind of effort that is required to transform the Malaysian manufacturing ecosystem to upgrade economic agents to be competitive in the age of IR4.0 systems. By 2030, there will be a fundamental paradigm shift in how technologies will be utilized in society and underpinning that will be a host of innovative and disruptive business models being brought to the fore. Currently, the local industrial landscape lacks a sufficient critical mass of key core capabilities to catalyse growth in order to compete globally. Thus, there needs to be more ecosystem level initiatives that focus on seeding IR4.0 growth leveraging on Quadruple Helix collaborations. While the triple helix is still central to such a transformation, there is also a need to shift towards the quadruple helix so that it engages all socioeconomic agents in the country, including the consumers and critical civil society organizations that push for the inclusive participation of the people.

The heavy investment required to develop the ecosystem could be sourced through a PPP framework enabled by tools like which STRAND, which is developing that to manage the involvement of multiple and diverse stakeholders whilst providing robust governance through a digital platform that should ensure competitive outcomes via transparency, timeliness, accountability, effective empowerment and equitable distribution of rewards. Tools like the SIM have the ability to coordinate, align, manage and extract synergistic value from national ecosystems.

Acknowledgement

The authors wish to thank all the relevant industry and government players, such as UMW, Spirit Aerosystems, CTRM, Ministry of International Trade and Industry, Malaysia Investment Development Agency, Malaysia External Trade Agency, SME Corporation Malaysia, the Selangor and Kedah State Economic Planning Unit to name a few for their cooperation and support throughout.

References

AmInvestment Bank (2019) UMW Holdings – AmInvest Research Report, *AmInvest*, Kuala Lumpur. Available at https://www.bursamarketplace.com/mkt/tools/research/ch=research&pg=research&ac=772031&bb=786470 [Verified 7 August 2021].

DW (2018) *Spotify: How a Swedish Startup Transformed the Music Industry*, DW, viewed 7 August 2021, https://www.dw.com/en/spotify-how-a-swedish-startup-transformed-the-music-industry/a-43230609.

Fox, M. (2020) Tesla Overtakes Toyota to Become Most Valuable Automaker Car in the World (TSLA, TM), *Markets Insider*, viewed 7 August 2021, https://markets.businessinsider.com/news/stocks/tesla-stock-surpass-toyota-most-valuable-automaker-world-market-cap-2020-7-1029359765.

Gerschenkron, A. (1062) *Backwardness in Historical Perspective*, Cambridge: Belknap Press.

Haridy, R. (2019) *Netflix vs. Cinema: How a Disruptive Streaming Service Declared War on Hollywood*, New Atlas, viewed 7 August 2021, https://newatlas.com/home-entertainment/netflix-disruptive-streaming-hollywood-cinema-exhibition-war/.

Kaldor, N (1967). *Strategic Factors in Economic Development*, Ithaca: Cornell University Press.

Manufacturing Today (2020) *3D Printing and the Supply-Chain*, *Manufacturing Today*, viewed 7 August 2021, https://www.manufacturingtodayindia.com/sectors/8896-3d-printing-and-the-supply-chain.

MARA Aerospace and Technologies Sdn. Bhd (2017) *Stage 2: Toulouse Accelerator Programme Report*, Selangor: BDEC Resources Malaysia.

MARA Aerospace and Technologies Sdn. Bhd., Strand Aerospace Malaysia Sdn. Bhd (2017) *UMW High Value Manufacturing Park, Serendah Blueprint*, Selangor: UMW Land Sdn. Bhd.

MARA Aerospace and Technologies Sdn. Bhd., Strand Aerospace Malaysia Sdn. Bhd (2019a) *Transforming Selangor into A Regional Smart State Benchmarking Report*, Report No. 4, Kuala Lumpur: MARA Aerospace and Technologies Sdn. Bhd.

MARA Aerospace and Technologies Sdn. Bhd., Strand Aerospace Malaysia Sdn. Bhd (2019b) *Transforming Selangor Into A Regional Smart State Industry Theory Framework Report*, Report No. 5, Kuala Lumpur: MARA Aerospace and Technologies Sdn. Bhd.

McKinsey & Company (2021) *COVID-19: Implications for Business*, McKinsey & Company, viewed 7 August 2021, https://www.mckinsey.com/business-functions/risk/our-insights/covid-19-implications-for-business#.

Plummer, R. (2020) Coronavirus: Six Industries Crying out for Help, *BBC News*, viewed 7 August 2021, https://www.bbc.com/news/business-52200386.

Professional Development Centre (PDC) (2014) *Manual*, 1st ed. Kuala Lumpur, WP: MARA Aerospace and Technologies Sdn. Bhd.

Rasiah, R. (2019) Building Networks to Harness Innovation Synergies: Towards an Open Systems Approach to Sustainable Development, *Journal of Open Innovation: Technology, Markets and Complexity*, 5, 70.

Rasiah, R. (2020a) Industrial Hubs and Industrialization in Malaysia, Arkebe, O. & Lin, J. (eds), *Oxford Handbook of Industrial Hubs*, Oxford: Oxford University Press.

Rasiah, R. (2020b) Industrial Policy and Industrialization in Southeast Asia, Arkebe, O., Cramer, C., Chang, H.J. & Kozul-Wright, R. (eds), *Oxford Handbook of Industrial Policy*, Oxford: Oxford University Press.

Rasiah, R., Gopi Krishnan, V., and Azleen, O.R. (eds) (2022) *Financing Sustainability with East Asian Experiences*, Kuala Lumpur: Universiti Malaya Press.

Reinert, E. (2007) *How the Rich Became Rich … and Why the Poor Stay Poor*, London: Anthem Press.

Schumpeter, J.A. (1934) *Theory of Economic Development*, Cambridge: Harvard University Press.

Schumpeter, J.A. (1942) *Capitalism, Socialism and Democracy*, New York: Harper and Roe Publishers.

Strand Aerospace Malaysia Sdn. Bhd (2016) *UMW M&E Roadmap and Strategy Report*. Report No. 02–04, Selangor: UMW M&E Sdn. Bhd.

Tan, T., Sivanandam, H. and Rahim, R. (2020) Over 30k SMEs Have Shuttered since the Beginning of MCO, Parliament Told, *The Star*, viewed 7 August 2021, https://www.thestar.com.my/news/nation/2020/11/09/over-50k-smes-have-shuttered-since-the-beginning-of-mco-parliament-told.

UNESCO (2021) *UNESCO Science Report 2021: The Race Against Time for Smarter Development*, Paris: United Nations, Educational, Scientific, and Cultural Organization (UNESCO).

United Nations (2016) *Sustainable Development Goals Report 2016*, Geneva: United Nations, downloaded on May 23, 2022 from https://unstats.un.org/sdgs/report/2016/

University of Malaya, National University of Malaysia, MARA Aerospace and Technologies Sdn. Bhd. Aerospace Malaysia Innovation Centre (2014) *R&T Cooperation Agreement For Virtual Reality Immersive System Training For Aerospace Manufacturing*, Kuala Lumpur. Unpublished manuscript.

13 Government Initiatives to Promote Adoption of IR4.0 Technologies in Manufacturing

Abdul Latif and Saliza Saari

Introduction

Despite continued calls for raising the role of markets as a key institution in the promotion of cutting-edge technologies, it is largely accepted that the government continues to play an important role to ensure that its spread provides strong synergies for economic upgrading among the developing economies. In light of that, the government launched the Malaysia Productivity Blueprint (MPB) in 2017, aimed to boost Malaysia's economic growth and raise the prosperity of the rakyat. For Malaysia's recovery journey, it can be escalated by focusing on the five key drivers for productivity growth as envisioned and detailed in MPB.

Talent, technology, incentive, business environment, and productivity mindset are the key drivers to boost productivity amidst the impact of the COVID-19 pandemic. A deep dive into these key drivers, given the present global and domestic economic environment and critical issues and challenges at hand, paves a clearer path for implementing solutions and recommendations, which in the end will have a direct impact on the country. Quick wins and long-term solutions based on the key drivers can be formulated to accelerate economic growth faster, better, and sooner.

The challenges in these five core areas have been compounded by the worst economic crisis in a decade due to the COVID-19 pandemic. Issues were already looming pre-pandemic mainly attributed by disruptive technology and rapid globalisation. Repercussions from the COVID-19 crisis call for a more comprehensive look into the issues and barriers impeding productivity, progress, and performance to reassess the situation as business is no longer as usual.

The MPB highlights the need for productivity to be addressed holistically at all levels to ensure a systemic change across the economy, which is a departure from previous fragmented efforts to raise productivity. A strong coordination and governance mechanism has been established to ensure effective and transparent implementation of the Blueprint. The private sector will drive this productivity agenda in partnership with the government.

Productivity needs to be top of mind movement and embedded into day-to-day work culture. Industry needs to understand the impact that productivity will have on their bottom line and have access to a feasible method of tracking their

DOI: 10.4324/9781003367093-13

productivity. It is essential that government mechanisms encourage productivity (such as by ensuring that all incentives are linked to clear productivity outcomes) so that organisations adopt productivity tracking as the norm.

The Blueprint outlines immediate national-level priorities that require policy reform and the government's intervention. For sector-specific initiatives, a rollout in prioritised stages is proposed. At the enterprise level, the Blueprint describes the required expertise and support for enterprises to understand and tackle their productivity challenges on the ground.

The Blueprint implementation requires oversight through a robust governance model. There are four roles required going forward: strategic oversight, advisory, coordination, and monitoring as well as implementation. Strong coordination is critical to driving implementation on the ground, with rigorous programme management to ensure transparency and accountability. Sector Productivity Nexus play a key role in supporting institutions on the ground, simultaneously improving the visibility of the implementation progress.

Theoretical Considerations

While the world continues to witness enhanced roles played by market forces, there are a number of reasons why the government is still considered a central organisation to ensure that the critical regulatory instruments are in place to prevent market failures. The outbreak of the global financial crisis in 2007–2008 from arguably the most developed market economy, i.e., the United States, is a classic example of a catastrophic market failure that arose largely because the Fed representing the government had allowed markets to lead the development and diffusion of new technological instruments with deregulatory initiatives (Stiglitz, 2010; Krugman, 2009). Because Malaysia is still a developing economy, still mired in the processes of transition from medium value-added to high value-added economic activities, the need for government instruments to support firms' operations is critical owing to the need for shielding the economy from market failures.

Institutions are the key moulders of the conduct of economic agents (Nelson and Winter, 1982; Coase, 1992; North, 1994; Rasiah, 2011), and hence, are central to spreading its take up among both public and private individuals, firms and organisations. The government's regulatory instruments, both rewards and penalties, are a key influence in the adoption of the Fourth Industrial Revolution (IR4.0) technologies (see also Rasiah, 2019). A blend of both formal and informal institutions play critical roles in raising the take up IR4.0 technologies, governments are often the initiators among poor economies owing to a lack of a critical mass of private agents to offer the capital investments, as well as the lack of private interests to take up voluntary and un-rewardable activities associated with transactions where public returns exceed private returns.

In doing so, governments not only enacted regulations to attract businesses into risky and uncertain fields but also offer the initial spur for firms to enter new areas of business and to foray into frontier R&D (Rasiah, 2020). Also, governments also launch organisations that are necessary to solve collective action problems, which are sometimes eventually privatised once they mature. A considerable number of industrial technical research institutes (ITRIs) in Taiwan that began in 1974 have increasingly spread into private operations with partnerships with firms (Tsai & Cheng, 2006). The R&D consortium in Taiwan is one example (Mathews, 2002). State-level networks between government agencies and private firms at industrial parks have also mushroomed in Europe, the United States, and Japan (Best, 2018).

It is also imperative for countries (such as Malaysia) which is facing serious labour shortages to quicken the diffusion of IR4.0 technologies to reduce its dependence on less skilled foreign labour. It is through such mechanisms that Taiwan has managed to bring back significant amount of manufacturing and agriculture from China, (which started aggressively following President Trump's efforts to raise tariffs on goods imported from China since 2017) (Rasiah, Low and Nurliana, 2023). Japan took the same initiative in 2020 when offering USD 2.2 billion to induce its automotive firms to relocate from China to Japan and Southeast Asia (Denyer, 2020).

Multifactor Productivity (MFP) is identified as one of the drivers for robust productivity growth. The progress of an organisation's MFP is greatly influenced by technology. Adoption of more advance technologies optimises MFP, which in turn leads to increased productivity level. Within this scope, digital technology has marked its place in workplaces in streamlining processes and operations more efficiently.

Past literature has indicated a strong correlation between the application of digital technology and increased productivity level. Harvard Business Review (HBR) reported that "the most digital companies see outsized growth in productivity and profit margin". Three main areas in digital technology which impact business performance and productivity are digital assets, usage, and workers. Simply put, providing digital tools to employees to perform tasks enhances their productivity.

Technology-induced productivity and performance are mainly attributed to efficient and optimal use of firms' resources, energy, expertise, labour, and time. It is the smarter way of getting the job done. In 2020, when the COVID-19 pandemic led the government to impose the Movement Control Order (MCO) for the first time, organisations shifted from conventional operations to digital technology to deliver programmes, such as training, conferences, and seminars virtually. Online programmes helped save cost and resources by over 50 percent in comparison to the conventional methods. Outreach programmes reached wider coverage, while the digital programmes eliminated the need to travel. While digitalisation did remove the benefits of physical feel that often come with interactive sessions, its forced introduction caused by the COVID-19 pandemic offered MPC the opportunity to quicken its integration with the inevitable shift to the next epoch of technological regimes.

Given such a critical role played by governments to promote the proliferation of new technologies, it is imperative that similar developments should occur in Malaysia if the country is to strengthen the ecosystem embedding firms to make the transition to the adoption of IR4.0 technologies, which is also associated with increasing incremental and radical innovations to catch up with firms in the developed countries. Consequently, this chapter examines the role played by the Malaysian government in general, but the Ministry of International Trade and Industry (MITI) to strengthen the embedding ecosystem in the country to support firms' absorption of IR4.0 technologies.

Promotion of Digital Transformation

The government, via its Economic Planning Unit (EPU) has launched MyDigital blueprint that lays the roadmap to achieve the country's grand vision to become a regional leader in the digital economy and attain an inclusive, responsible, and sustainable socioeconomic development. The blueprint consists of six major thrusts that map out 22 strategies, 48 national initiatives, and 28 sectoral initiatives. MyDigital blueprint comprises action plans which adopted a whole-of-nation approach to complement the Twelfth Malaysia Plan (12MP) and the Shared Prosperity Vision 2030.

MyDigital outlines the plans to accelerate Malaysia's progress as a technologically advanced economy, through the Malaysia Digital Economy Blueprint. This will chart the path to strategically position Malaysia as a competitive force in this new era. This blueprint sets out a combination of initiatives and targets across three phases of implementation until the year 2030; Some highlights include:

- The digital economy is expected to contribute 22.6 percent to the country's GDP by 2025. At the same time, the plan also aims to open 500,000 job opportunities in the digital economy in 2025.
- The government will also encourage 875,000 micro, small and medium enterprises to accept the use of e-commerce.
- Initiatives under the auspices of MyDigital can also catalyse 5,000 start-up companies or start-ups in the next five years.
- MyDigital will also be a starting point to attract new investments in the digital sector of RM70bn from within and outside the country.
- The government is targeting the level of productivity of the economic sector to increase by 30 percent by 2030 compared to 2021.
- All agencies in the public sector will provide cashless transaction facilities as the main choice by 2022.

The National IR4 Policy is the policy supported by the Malaysia Digital Economy Blueprint. The Policy and the Blueprint will act as guiding documents for the rakyat to leverage the potential of IR4.0. A governance structure led by the National Digital Economy and IR4.0 Council has also been established to drive and ensure effective implementation of initiatives which cut across various ministries and agencies. The government recognises the benefits that industries could enjoy in harnessing IR4.0 towards achieving Malaysia's long-term goals outlined in our national policies.

The government continues to formulate and implement policies and initiatives to create a more conducive environment and provide opportunities for the continued growth of the digital economy. The launching of the IR4.0 Policy in October 2018 marks a significant milestone in Malaysia's bid to be on par with the developed nations in the use of IR4.0 (Malaysia, 2018). Industries and business communities are expected to quickly innovate and transform to embrace IR4.0 by utilising various intelligent digital technologies to uplift productivity. Their failure may result in stagnating productivity and negative effects on the economy, society, and environment.

Malaysia has formulated and implemented the right policy, positive incentives, and conducive ecosystem to drive the transformation in the push to become an innovation-driven economy. This has resulted in Malaysia being ranked second for Global Competitiveness among ASEAN Countries by the World Economic Forum's Global Competitiveness Report. Malaysia's performance in key global indices has reflected the effectiveness of national policies in developing the digital economy.

Technology advances leads economy towards sustainable growth. Thus, technology adoption of a country provides an essential basis in explaining the country's competitiveness and where it ranks among other countries. The use of internet and technology advancement contributes to the rapid growth of data, which is the future commodity. Nevertheless, countries risk creating digital divide if the response to digitalisation is not managed well. For Malaysia to stay competitive and relevant, digitalisation should be embraced and opportunities arising from this trend should be seized.

Since the beginning of COVID-19 pandemic crisis, businesses that do not keep up with the current technology and digital trends and do not take preemptive measures to apply digitalisation as part of their business's competitive edge will not survive the market. Certainly, Malaysia is gaining a huge awareness, as exemplified by its efforts in adopting digital transformation (DT) in various industries. The COVID-19 pandemic has catalysed the rapid transformations in the manufacturing sectors.

IR4.0, also known as "smart manufacturing", is the cutting-edge development in the digitisation and industrial automation of manufacturing processes. It incorporates the most advanced and innovative technologies such as industrial internet of things (IIoT), additive manufacturing (AM), industrial artificial intelligence (IAI), big data analytics (BDA), autonomous robot, cloud computing, cyber security, augmented reality (AR), simulation, and advanced materials.

IR4.0 covers the entire end-to-end value chain that starts from suppliers, procurement, production, design, logistics to sales to achieve higher productivity and output flexibility. For example, the usage of cyber-physical ecosystem enables businesses to monitor the physical processes of the factory, which in turn enables them to make intelligent decisions. Physical systems make up part of the sensory mechanism of the IIoT for communicating, analysing, cooperating, and interacting with each other and with humans in near real time via wireless web-platforms.

In a nutshell, businesses can reduce operational cost as IR4.0 emphasises on streamlining overall business operations by minimising wastage or storage, enhancing supervisory processes and maintenance of machinery, at the same time instilling streamlined security and safety efficiencies. Despite the government's continuous and nation-wide investment and initiatives to modernise and advance the state of DT, technology adoption rate remains low. Many enterprises still prefer the traditional ways of doing business while Malaysia has consistently strived to become an innovation-driven economy.

Digital adoption within industries is still in its infancy as industrial development indicates sluggish progression and less aggression by comparison to the neighbouring countries. Malaysia's 907,065 small and medium enterprises (SMEs) are living in very uncertain times. SMEs are almost impossible for them to predict what will come next. Businesses across the country have barely hung on after the arrival and persistence of COVID-19. The digitalisation and automation journey are able to push businesses to the speedy recovery from COVID-19 pandemic crisis.

Overcoming Barriers to IR4.0 Adoption

In general, Malaysia is still struggling to adopt IR4.0, and many businesses are stuck at Industry and Industry 3.0, in terms of manufacturing technology. The common challenges faced by many manufacturing businesses in implementing IR4.0 are as follows:

i   Uncertainty and hesitation to modernise infrastructure and their legacy system.
ii Lack of capabilities in converting huge amounts of data collected into useful business insights and using these insights to improve operational efficiencies.
iii Lack of data security and skillsets needed in digitising their processes.

Due to these challenges, many manufacturers still rely on low-cost labour for example foreign workers and are hesitant to invest in innovative automation technologies.

Manufacturers need to move with the market, but movement will only happen when it is justified by a viable business case. The Malaysian government, industry leaders, private sectors and the local academia need to work together to prove, with practical and pragmatic business cases, that DT is fundamental for an organisation to enhance customer service and experience.

MITI and its agencies, in collaboration with other relevant Ministries and Agencies, are undertaking various outreach programmes to increase public, industry, academia, and training institute's awareness on IR4.0. Targeted incentives and funding are provided to promote the adoption of IR4.0 under the Technical Working Groups (TWG) for Incentives and Funding.

The private sector industry is also being urged to invest in their own digitalisation efforts. Many are already putting in effort in implementing IR4.0; however, more concerted efforts from industry leaders need to be seen. There are organisations from the electrical and electronics (E&E), aerospace, and the automotive sectors that are more advanced in terms of IR4.0 adoption. These success stories by industry captains can then be set as a benchmark for others to emulate moving forward.

Engaging a DT consultant can enable greater understanding of the context and better insight into enterprise's operation as well as connect the company with the niche specialists and technology partners that are the best fit for the company's business needs. It is very important to engage a consultant with a combination of technical knowledge, business knowledge, change management skills, and implementation experience.

The DT consultant plays an important role to assist the company to create a clear strategy and roadmap of enterprise's DT journey. The DT consultant is able to identify any gaps in the enterprise's DT process as well as being an integral part of the DT process. A DT consultant understands both the current traditional and the desired digital side of the business and can assist with change management in the company by designing a tailored strategy. A good DT consultant can formulate cost-effective solutions, solve complex business challenges, and drive continuous improvement in the company.

Stimulating the Diffusion of IR4.0

Pacing automation and robotics into the seams of a manufacturing foundation is crucial in the elimination of human errors and enables businesses to refocus their effort in areas which drives revenue for the company. It drives and encourages greater enhancements in manufacturing processes and sets local market benchmarks in attaining quality production output, efficient practice, and better profit-gaining opportunities for businesses.

Rising global competition has driven the intensification of technological change in the manufacturing sector, but the proliferation of IR4.0 has imposed epochal changes that necessitate revolutionary transformations for firms to compete in the new economy (see Yeap & Rasiah, 2023). Businesses have been changing dramatically over a short period of time and exposed to uncertainty, complexity, and risks that influenced by technological innovation. Additionally, IR4.0, which is rapidly approaching, has created massive, unprecedented change and barriers across the manufacturing sector. Employment disruptions, high implementation cost, organisational and process changes, security and privacy issues, regulatory compliance issues, and lack of data management are few barriers that are required to cope up with the upcoming emerging technologies. The growing global competition especially adopting IR4.0 into the manufacturing sector has mounting challenges to the business world.

National Policy on Industry 4.0 Technologies for Manufacturing

National Policy on Industry 4.0 is a national policy that aims to change or transform the manufacturing sector and related services through 13 strategies and 38 action plans. The policy outlines five national strategies, including finance, infrastructure, legislation, skills and talent, and innovations, with the aim of attracting stakeholders to Industry 4.0 technologies and processes while also increasing Malaysia's attractiveness as a preferred manufacturing location. This strategy is expected to build the right environment for Industry 4.0 to take hold, as well as align current and future growth initiatives, resulting in a comprehensive and rapid transformation of Malaysia's manufacturing capabilities.

The manufacturing industry that made up of a large number of SMEs, approximately 99 percent of the manufacturing firm, has contributed to 22 percent of Malaysian gross domestic product (GDP) in the past five years. The period of this policy is effective from 2018 to 2025, demonstrating Malaysia's contribution to the United Nation's Sustainable Development Goals (SDGs), specifically in support of Goal #9 Industry, innovation and infrastructure and Goal #12 Responsible Consumption and Production. This policy consists of three visions in making Malaysia as:

* strategic partner for smart manufacturing and related services in the Asia-Pacific;
* primary destination for investment in the high technology industry; and
* a total solutions provider for cutting-edge technology.

Recent data of Global Innovation Index 2020 by Cornell University shows that Malaysia has risen to 33rd place, up from 35th place previously. To achieve the goals, the enabling technologies introduced and nine described by this policy known as artificial intelligence, big data analytics, augmented reality, additive manufacturing, cybersecurity, simulation, advanced material, system integration, autonomous robotics, internet of things (IoT), and cloud computing introduce a new dimension to the manufacturing landscape, resulting in a significant increase in industrial productivity.

The foundation of the nation to achieve a fully developed country is by transforming itself into export-led growth. The importance of the export-led growth to Malaysia is directly linked to the development of its manufacturing industry. The innovation and creativity of the manufacturing industry in Malaysia to cater the demand of other countries open external trades that heavily contribute to its economic growth.

According to McKinsey Industry 4.0 ASEAN Survey 2017, the percentage of IR4.0 awareness among companies still low with the implementation of IR4.0-related technologies in their business. The survey highlighted top-five reasons that holding back the IR4.0 implementation. Firstly, problems to define clear business plan. Secondly, a very huge data is not integrated across business units. Thirdly, there is no digital talent that able to execute the IR4.0 roadmap. Fourthly, organisations are very concerns on cyber security risks. Lastly, there is no coordination across business units.

Enterprise Resilience through Digitalisation and Innovation

Key to sustaining a business is productivity. It is about achieving maximum efficiency and effectiveness through creating higher value products to customers. As productivity needs to be embedded into day-to-day work culture, enterprises in manufacturing sectors need to know how to improve their business operations.

Digitalisation and innovation are crucial for enterprise to survive and sustain in this new norm in managing business. To spearhead the growth of the industry and adoption of technology in wider perspective, enterprises need to optimise the utilisation of digital in holistic manner coupled with continuous innovation and creativity. In 2020, a total of 11,322 organisations have benefited from various programmes organised by MPC. The programmes covered wide scope ranging from assessment, intervention, best practices documentation to sharing session.

Productivity Improvement Programmes

Productivity improvement programme (PIP) is a programme conducted by MPC to improve efficiency and effectiveness of production or services. The overall impact of PIP programme is increase productivity and competitiveness of the company. By participating in this programme, the company will implement a continuous improvement programme starting from the diagnostic process, training, intervention (identifying the root cause of the problem, improvement solutions, implementation of improvement solution and monitoring the new process), and subsequently measure the impact.

Upon completion, MPC will measure the outcome achieved by the enterprises such as productivity improvement, waste elimination, reject rate, delivery, cost saving, and customer satisfaction. The overall impact achieved by the enterprises in these programmes such as waste reduction came to about 60–90 percent by eliminating non-value- added processes in production or services, increase efficiency in the production time of a product or services by 50–70 percent, and quality improvement of 50–90 percent by reducing the reject rate of product damage and improving the quality of the product or service. The impact of the programme is cost and time saving, elimination of unnecessary wastes, delivery within the promised time and high-product quality. This outcome will lead to assist on increasing customer satisfaction and increasing organisational profitability with profit rate increased by 30–50 percent.

On a global scale, Malaysia has a reasonably good and competitive role in both manufacturing and technology use. Malaysia was ranked 17th out of 40 countries in the Global Manufacturing Competitiveness Index 2016 (by Deloitte Touche Tohmatsu). According to the report, Malaysia will rise four places to

13th by 2020. In view of technology and innovation, the Global Innovation Index 2017 (conducted by Cornell University, INSEAD, and WIPO) ranked Malaysia 35th out of 127 countries and eighth in Asia. Furthermore, the previous report by the World Economic Forum (WEF) and A.T. Kearney Readiness for the Future of Production Report 2018 shows that Malaysia is well-positioned to benefit from the future of Industry 4.0.

This report has highlighted Malaysia in the "Leader" quadrant. These are countries that have a "strong current output base" and are "well positioned for the future". It is also worth noting that Malaysia and China are the only two upper-middle-income countries in the "Leader" quadrant, with the rest coming from high-income countries, nonetheless, it is important to note that the majority of high-income countries have ambitious plans and are moving rapidly to implement them but this country still experienced this significant gap.

According to the WEF's report, the main drivers of development for Industry 4.0 are technology, human resources, global trade and networks, and institutional structures. Malaysia ranks 21st to 30th out of 100 countries in each of these drivers, emphasising priorities in technology, human resources, and institutional structures. Initiated by the strong report justification, Malaysia has designed the Industry4WRD Readiness Assessment programme to assess an organisation's level of readiness to transition to Industry 4.0. The main objectives of this program are to:

- create guidelines on the level of organisational readiness in the use of Industry 4.0 elements;
- identify areas of improvement in each dimension measured;
- recommend further action to improve efficiency and productivity; and
- develop a foundation for industry adoption Productivity Nexus Digital Initiatives.

MPC Productivity Nexus to Boost Manufacturing Initiatives

MPB has identified the challenges faced by different sectors and industries in Malaysia. The challengers provide a strong justification for establishing additional sector-specific initiatives to unlock the potential of productivity in different sectors and industries. Therefore, Sector Productivity Nexus that cover nine subsectors are led by industry champions from the sectors or industries. The nexus has been formed and led by industry champions to drive the implementation of the initiatives of the nexus.

Under Productivity Nexus, industry associations and enterprise champions are empowered to leverage on their strong connections and networks to engage with sector players. The objective of the nexus is to materialise the intended impacts of the MPB through well-designed implementation of the programmes. For the past years since its launch in 2017, these nine Productivity Nexus have organised programmes under the stated initiatives by utilising the public-private partnership governance model.

We examine below six government initiatives to promote the adoption of digitalisation under the Productivity Nexus and Manufacturing sub-sectors in Malaysia. Although it is still early to assess the impact of these initiatives, the direction has been set and MITI is eager to see the successful execution of these initiatives for maximum impact on manufacturing firms.

Agro-food Productivity Nexus

The agro-food sub-sector is undergoing a transformation driven by new technologies that enable this primary sector to uplift its productivity and profitability. The significant benefits of adopting this ever-evolving digital agriculture include lower consumption of water, nutrients, and fertiliser, better decision-making through data science, along with the less detrimental impact on the surrounding ecosystem, thus making farming business more cost-effective, innovative, and sustainable.

In this regard, Agro-food Productivity Nexus (AFPN) has developed a compendium of agro-food technology in 2020 that offers in-depth, reliable, and valuable information regarding up-to-date knowledge and existing technology in the market. It covers the sub-sector supply chain, namely production, post-harvest handling, processing, distribution, and retail. The foundation of this compendium also brings the upstream and downstream stakeholders of the agro-food value chain together in a single platform.

To date, there are about 618 visitors who have already reaped the benefits of rich information possessed by this database. Sufficient access to information through this compendium will empower the farmers to bring the full benefits of technology for supporting productive and sustainable farming practices. It will also help them to transform the existing land into more modern and competitive farms with the purpose of maintaining viable food production while improving the competitiveness of the entire agro-food value chain.

Chemical Productivity Nexus Digital Initiatives

Chemical Productivity Nexus (CPN) has successfully conducted the initial pilot project on Chemical Virtual Advisory Clinics (CHEM-VAC), which is a part of Business Virtual Advisory Services. Thirty virtual clinics were conducted last year involving ten chemical industrial experts act as advisor that provide assistant, advice, and recommendation to 30 chemical SMEs effected by the COVID-19 pandemic. More than 100 recommendations gather focusing on financial consultation, innovation business operation, supply chain management, and intelligence manufacturing solution, which intend to assist the Chemical industry players to recalibrate the industry for a stronger comeback once the impact of COVID-19 is contained.

Digital Productivity Nexus Digital Initiatives

The Digital Productivity Nexus (DPN) has taken lead to promote the effective use of digital technology through its "Go BIG with Digital" initiative under the

Malaysian Productivity Blueprint. Go B.I.G with Digital focuses on catalysing productivity growth for Breakthrough results, strengthening Integrity and empowering best practices and Good values to increase productivity growth or "Go B.I.G with Digital". This will be done through the adoption of IR4.0, producing stronger connectivity and innovation of technology.

Digital STARS Programme

In the world of MSMEs, being resilient to uphold the business can easily be distracted by employee performance issues. The profound challenges in dealing with staff turnover and acquiring high-quality employees may halt the effort towards digitalising the business operation in response to the COVID-19 pandemic. This alarming circumstance is why DPN has developed a Quadruple Helix Framework-Digital STARS Internship Framework, a digital platform to address a shortage of digital technology talents and strengthen the collaboration among industry, academia, government, and community.

Electrical and Electronics Productivity Nexus

The Electrical and Electronics Productivity Nexus (EEPN) has the charter of enhancing the competitiveness of the E&E industry, particularly in increasing the productivity, capability, and competency of the E&E subsector. One of EEPN's flagship initiatives is the Plugfest workshop that aims to accelerate Industry 4.0 among the local industries. This programme aims to train and reskill the workforce, especially those affected by the pandemic, to stay relevant in the industry. This programme consists of six to seven half-day hands-on workshop and the Proof-of-Concept (PoC) project within three to six weeks after the workshop.

Plugfest 2.0

In 2020, EEPN has kickstarted Plugfest 2.0 programme after successful implementation of the previous Plugfest 1.0 programme. Plugfest 2.0 programme generally focuses on harnessing the benefits of artificial intelligence-based (AI-based) machine vision system compared to Plugfest 1.0, which emphasises the application of the IoT. The primary objectives of Plugfest 2.0 are to promote AI-based machine vision technology, which is a critical element of Industry 4.0, and empower the participants to be able to embed machine vision system in their work environment.

EEPN, together with its technology partners, provides coaching to participants during the workshop session. The participants can test and use necessary AI-based machine vision system–related starter kits (e.g., Machine Vision System with Integrated high- performance Intel Core i7 processor, system memory, storage, camera, and deep learning software). Upon completing the workshop, participants will bring back the kits for carrying out the PoC project at their company. Forty-five engineers from 25 companies have participated in this programme from August to October 2020. EEPN is currently devising a plan to proliferate this programme to a larger scale.

Machinery and Equipment Productivity and Digitalisation Nexus

The Machinery and Equipment Productivity Nexus (MEPN) has conducted a Machinery & Equipment Virtual Advisory Clinic (MEVAC), with 43 companies during the MCO and provided recommendation reports to them to assist in mitigating the impact of COVID-19 and to further improve companies' productivity. The one-to-one consultations via an online platform were assisted by eight Industry Productivity Specialists as advisors. MEVAC has identified 123 issues and business concerns, and four programmes have been designed to address these issues namely Business Financial Dialogue Session, Productivity 1010, and Trade Coordination with MATRADE.

Realising the importance of digitalisation to improve productivity, the MEPN, with the support of the Malaysia Automation Technology Association (MATA) and the Malaysia Industry 4.0 System Integrator Association (Misi4.0), has developed Digitisation Self-diagnostic Tool and Digitisation Prioritisation Matrix to help manufacturing companies to strategise their individual digitisation plans and kick-start their digitisation journey.

This module encompassed a digitisation self-diagnostic tool and a virtual mentoring session. The programme, known as Productivity 1010, targeted providing assistance to 1000 manufacturing companies, with the ultimate output being an Individual Digitisation Blueprint, and with the help of 20 qualified industry experts as mentors. To date, 79 companies have used the digitisation self-diagnostic tool and 37 companies have assisted via virtual mentoring session.

Retail and Food and Beverage Productivity Nexus

The Retail and F&B Productivity Nexus (RFBPN) has successfully conducted 281 clinics for previous RFBVAC 2020, which involved 39 retail and food and beverage (F&B) industrial experts act as advisor to assist 281 retail and F&B SMEs effected by the COVID-19 pandemic. Area of advisory focus on branding and marketing strategy, supply chain management, business digitalisation, business operation and standard operating procedure (SOP), human resource management, and business funding which intend to assist the retail and F&B industry players to recalibrate the industry for a stronger comeback once the impact of COVID-10 is contained. With positive feedback from the industry, RFBPN feels this effort should be carry on gathering more business concerns and provided hands on guidance through direct recommendation from the industry expert. Future RFBPN initiatives can be derived from the outcome of this programme.

Certificate in Food and Beverage Operations

F&B enterprises recognise the importance of investing in and developing their human capital that motivates employees to provide high-quality customer service and improve efficiency. Adequate training and growth opportunities can significantly boost employee satisfaction and loyalty. The combination of high service efficiency and low turnover can potentially increase the subsector productivity. Highly skilled employees operating the F&B enterprises is also crucial to cater for the uprising expectation of the customers resulting from extensive utilisation of social media.

Therefore, the RFBPN has developed a certification programme in F&B operations to upraise the standard of service delivery in the F&B industry. This programme covers eight days of skill-based and management competency aspects comprising five modules:

- Food and Beverage Service Operations
- Service Excellence
- Food Safety Programme
- Food and Beverage Management
- Effective Leadership Skills

By completing this certification programme, the participants will be able to raise their service standards to be at par with international standard, understand the culture of service excellence and obtain knowledge on food and beverage operation in a range of establishments.

Conclusions

Given the critical role governments play in addressing market failure in promoting manufacturing, this chapter examined the important initiatives undertaken by the Malaysian government to promote the diffusion of IR4.0 technologies in the sector. Although the IR4.0 and digitalisation policy blueprints were only launched in 2019 and 2020, respectively, the government had already initiated these developments through ad hoc policies from 2010. Following the launching of the blueprints and the Twelfth Malaysia Plan, the government consolidated these programmes and started to offer the incentives and the institutional development to support the adoption of IR4.0 technologies.

MITI in particular has embarked on both monitoring and promoting the adoption of IR4.0 technologies in Malaysia. A wide range of programmes have been adopted by MITI to strengthen the ecosystem to quicken the absorption of IR4.0 technologies in manufacturing, which has been driven by both interactive sessions with end-users and adaptation of best practices across the world. In the process of its facilitative role, MITI has supported not only the introduction of the new requisite skills that are essential to stimulate the growth of digitalisation and IR4.0 technologies but also the certification for them. Indeed, the government did not only earmark, but has also been actively interacting with both multinational companies and domestic small and medium companies to quicken the absorption of IR4.0 technologies in the country.

References

Best, M.H. (2018) *How Growth Really Happens: The Making of Economic Miracles through Production, Governance, and Skills*, Princeton: Princeton University Press.

Coase, R. (1992) The Institutional Structure of Production, *American Economic Review*, 82(4): 713–719.

Denyer, S. (2020) Japan Helps 87 Companies to Break from China After Pandemic Exposed Overreliance, *Washington Post*, July 21, accessed on June 28, 2022 from https://www.washingtonpost.com/world/asia_pacific/japan-helps-87-companies-to-exit-china-after-pandemic-exposed-overreliance/2020/07/21/4889abd2-cb2f-11ea-99b0-8426e26d203b_story.html.

Krugman, P. (2009). *The Return of Depression Economics and the Crisis of 2008*, New York: W.W. Norton.

Malaysia (2018) *National Policy on Industry 4.0*, Ministry of International Trade and Industry, accessed on August 26 from https://www.miti.gov.my/miti/resources/National%20Policy%20on%20Industry%204.0/Industry4WRD_Final.pdf.

Malaysia (2020) *Malaysia Digital Economy Blueprint*, Economic Planning Unit (EPU), accessed on August 26 from https://www.epu.gov.my/sites/default/files/2021-02/malaysia-digital-economy-blueprint.pdf

Mathews, J. (2002) The Origins and Dynamics of Taiwan's R&D Consortia, *Research Policy*, 31(4): 633–651.

Nelson, R.R. & Winter, S.G. (1982) *An Evolutionary Theory of Economic Change*, Cambridge: Harvard University Press.

North, D.C (1994) Economic Performance Through Time, *American Economic Review*, 84(3): 359–368.

Rasiah, R. (2011) The Role of Institutions and Linkages in Learning and Innovation, *Institutions and Economies*, 3(2): 165–172

Rasiah, R. (2019) Building Networks to Harness Innovation Synergies: Towards an Open Systems Approach to Sustainable Development, *Journal of Open Innovation: Technology, Markets and Complexity*, 5, 70

Rasiah, R. (2020) Industrial Policy and Industrialization in Southeast Asia, Arkebe, O., Cramer, C., Chang, H.J., and Kozul-Wright, R. (eds), *Oxford Handbook of Industrial Policy*, Oxford: Oxford University Press.

Rasiah, R., Low, S.W.Y. & Nurliana, K. (eds) (2023) *Digitalization and Development: The Ecosystem for Promoting Industrial Revolution 4.0 Technologies in Malaysia* (cross referenced from same book).

Stiglitz, J. (2010). *Freefall: America, Free Markets, and the Sinking of the World Economy*, New York: WW. Norton.

Tsai, T. & Cheng, B.S. (2006) *The Silicon Dragon: High-Tech Industry in Taiwan*, Cheltenham: Edward Elgar.

Yeap, K.L. & Rasiah, R. (2023) Diffusion of IR 4.0 Technologies in Electronics Manufacturing: The Role of the Embedding Ecosystem, Rasiah, R., Low, S.W.Y. & Nurliana, K. (eds) (2023) *Digitalization and Development: The Ecosystem for Promoting Industrial Revolution 4.0 Technologies in Malaysia* (cross referenced from same book).

14 Addressing Cybersecurity Issues

Maslina Daud and Rajah Rasiah

Introduction

The adoption of digital technologies has become a national agenda in Malaysia, where the Fourth Industrial Revolution (IR4.0) technologies are increasingly promoted to contribute positively to the digital economy (Economic Planning Unit, 2021a). The complexity of IR4.0 with various advanced technologies that introduce new security threats demands organizations to have a change of mindset and strategically plan on risk mitigation while doing the transformation. It involves not only challenges in investment in the technologies but also the capacity and capability of the workforce in ensuring that deployment is successful to eventually meet the business's objectives (Rasiah, 2019).

Regardless of the components and features that characterize IR4.0 systems, data need to be protected throughout its lifecycle, which can be challenging due to the massive amount of data involved, its heterogeneity, variety of sources, and the nature of its storage and transmission in multiple platforms from physical to the cloud environment. Despite its complexities, the data owners have the responsibility to ensure data security objectives in the IR4.0 system are met with protection and preservation of confidentiality, integrity, and availability. In doing so, the government becomes a central organization in providing the regulatory environment to install and manage a safe cybersecurity system for the active deployment and maintenance of IR4.0 technologies in individual countries. However, given the need for cooperation among the actors and consumers involved, such a regulatory body must have strong representation from the critical stakeholders.

Consequently, this chapter seeks to evaluate the components of the embedding ecosystem to assess the developments and readiness of the cybersecurity environment in Malaysia. The rest of the chapter is organized as follows. The next section looks at the national cybersecurity policies in Malaysia, which is followed by a discussion on cybersecurity issues arising from the main technology domains in the country. The subsequent section addresses the cybersecurity concerns associated with the deployment of IR4.0 technologies in Malaysia. The final section finishes with the conclusions.

DOI: 10.4324/9781003367093-14

National Policies on Cybersecurity

It is always difficult to see clear trends on cybersecurity breaches owing to reporting problems as several ends up not being reported. Figure 14.1 shows the total reported cybersecurity incidents in Malaysia. The reported incidents rose initially from 2006 until 2011, and sharply from 2009 until 2011 before showing a trend fall until 2017. It rose again in 2018 and plateaued since 2021. Within the breaches, fraud has continued to show a massive rise over the period 2006–2021 (Figure 14.2).

According to IoT analytics, the connected IoT devices globally are expected to grow at an annual average rate of 9 percent to reach 12.3 billion active connections. Malaysia's Digital Economy Blueprint is the umbrella policy that spearheads the country's move into the digital economy. This blueprint underlined a number of strategies to usher in the digital economy in Malaysia through a secure ecosystem with cybersecurity strategies drawn up to complement the policies. Two specific policies in the blueprint address IR4 proliferation, Industry-4WRD focuses on the manufacturing sector (Ministry of International Trade and Industry, 2018) and the IR4.0 policy encompasses almost every industry and all aspects of human life (Economic Planning Unit, 2021b).

The three sectors that were deliberated in the National Digital Blueprint are manufacturing, agriculture and services which are the focus of this chapter.

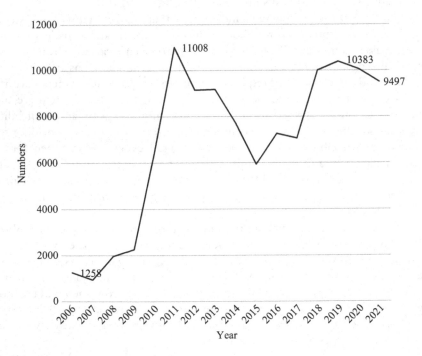

Figure 14.1 Total Reported Cybersecurity Breaches, Malaysia, 2006–2021.

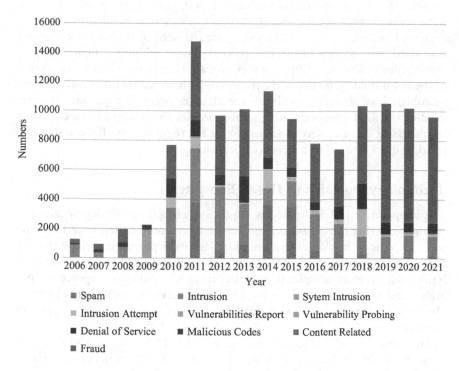

Figure 14.2 Cybersecurity Breaches Report by Category, 2006–2021.

Specifically, Thrust six focuses on cybersecurity that aims to create a trusted, secure and ethical digital environment (Economic Planning Unit, 2021a). Four strategies were formulated under this thrust, viz., first, strengthening safety and ethics in digital activities, second, enhancing institutions' commitment to data protection and privacy, thirdly, cross-border data transfer improvements, and finally, increasing cybersecurity uptake among businesses.

The smartification of manufacturing through the introduction of IR4.0 programmes requires improvements in data integrity, security standards and compliance to provide a seamless integration platform to raise problem-free and continuous upgrading of coordination between different actors in manufacturing value chains (Ministry of International Trade and Industry, 2018). The National IR4.0 policy emphasizes the legal framework that governs personal data management and cybersecurity to boost trust in online activities (Economic Planning Unit, 2021b). The national IR4.0 policy was established to enhance Malaysia's existing cybersecurity framework by incorporating safeguards for the smooth implementation and operationalization of such technologies across the public sector, with a focus on internet of things (IoT) (Economic Planning Unit, 2021b). In addition, cybersecurity strategies under the Malaysia Cyber Security

Strategy (MCSS) were crafted to embrace IR4.0, but these strategies are meant to be adopted only until the year 2024, which is a short-term initiative (National Security Council, 2020). It will also require the need to consider open innovation systems to address intellectual property issues without compromising on cybersecurity (Rasiah, 2019). The focus on short-term measures takes account of the fast unfolding nature of the technologies involved that demands frequent recalibration of strategies to absorb the changes necessitated. Thus, in spearheading the digital economy, long-term strategies are expected to be developed in addressing cybersecurity concerns so that the regulatory bodies can stay agile to adapt quickly to adjust to technological change.

Technology Domains in IR4.0 Ecosystems

The adoption of IR4 technologies requires organizations to deploy advanced technologies that the organizations can use to customize their operating environment. The technologies that are pillars for its deployment include the IoT, robotic process automation (RPA), artificial intelligence (AI), and big data analytics, which is more complex in manufacturing where the convergence of operational technology (OT) and conventional information technology (IT) systems already face a challenging operational environment due to security threats and vulnerabilities.

Apart from security threats involving traditional IT systems, IR4.0 encompasses technologies that introduce security risks due to its distinct characteristics. The following sub-sections will discuss some of the threats involving innovative technologies deployed in IR4.0 systems, including IoT, RPA, AI, and big data.

Internet of Things

The prevalence of IoT devices or sensors demands increased interconnectivities to allow data to be gathered and transmitted not only to other devices but also to other systems that process data. According to IoT analytics, the connected IoT devices globally are expected to grow at an annual average rate of 9 percent to reach 12.3 billion active connections and 27 billion active connections by 2025 (Sinha, 2021). IoT is meant for consumers' convenience in the domestic environment, while industrial internet of thing (IIoT) refers to an expansion of the usage of IoT devices linked and connected to a network with sensors and other equipment that is typically applied in industrial sectors, such as in smart manufacturing (Javaid et al., 2021).

IoT devices extended into manufacturing to provide remote access, effective monitoring, and better predictive analytics (Javaid et al., 2021), where the integration of information processing facilities and physical devices generates data through activities related to real-time sensing and communication. It is crucial to understand how the collected and stored data is preserved throughout this process. IoT devices are known to lack embedded security features due to its design, which focus on convenience rather than security features. IoT devices are normally not well-equipped with authentication or encryption features that provide data confidentiality capability. Worst still, users are often ignorant that not changing factory or default passwords exposes them to easy attacks. In the case of the Mirai attack, the malware infection turned into a bot that infected other similar devices (Krebs, 2016).

Most IoT devices are highly constrained with limited memory and processing power that do not offer security patches and update capabilities, which lead to security and privacy issues (Obaidat et al., 2020). One of the most noticeable aspects of IR4 is the convergence of technologies through the integration of processes. In a cps environment, components with intelligence embedded in sensors may be exposed to perpetrators if they are not configured securely. Since processes will be automated allowing the device with embedded intelligence to perform their tasks, its security will be overruled by the efficiency of the usage of the devices. The situation will be riskier when data produced by these devices are sent to the cloud environment.

Artificial Intelligence

For organizations where their businesses are data-driven and using an AI model is critical, security should be their top priority. According to Dror (2022), two main assets of the organizations are big data that is collected to learn and train for prediction purposes, while another is the data model arising from the training of the AI. Thus, these are the main assets that need to be protected to ensure they remain at the competitive edge.

AI is frequently deployed for monitoring and surveillance of systems where potential threats can be flagged when a defined threshold is met. Traffic patterns can be analysed to detect an activity that characterizes an attack (Kuzlu et al., 2021). Due to its capabilities, AI is considered one of the promising methods that are able to address cybersecurity threats. In IoT or Industrial environment, AI protects the systems by commonly used for intrusion detection, where it will notify possible attacks through anomaly detection on network activities in the systems. On the side of cyber attackers, AI has been increasingly used to thwart algorithms that perform this function (Kuzlu et al., 2021). Cyber perpetrators also use malicious AI to aid cyber-attacks or make the AI attack its own system.

Security threats on AI are mainly related to inputs of AI systems which can have adverse impacts on the integrity of the AI system. The wrong output will be produced by an AI system through data alteration, dataset poisoning, algorithm poisoning and model poisoning attacks (Kuzlu et al., 2021). In certain areas where AI is used, human intervention is still required. For instance, in doing software testing, a software specialist needs to review the findings of the software testing to ensure its accuracy. In a similar analogy, a medical doctor still needs to review and interpret lab diagnosis results. AI could also pose privacy risks when a comprehensive data collection and monitoring process outweighs the benefits the consumers may get. Through biometric profiles generated through these activities, consumers will feel threatened by this type of surveillance (Nematollahi & Khorasani, n.d.).

Robotics Process Automation

Moving towards IR4.0, RPA technology is deployed in businesses with the aim to improve efficiency. RPA programme allows a bot to be created to automate a manual repetitive process without human involvement. Despite the functions demonstrated by RPA, there are several security concerns that organizations need to be aware of. In IR4.0, full automation of processes is deployed, and the use of robotics particularly in manufacturing allows critical processes to be automated.

However, misconfiguration of a robot presents risks of stolen data, damaged end products and process disruptions to organizations and businesses (Gautam, 2022). Thus, Gautam (2022) further suggests countermeasures including; multifactor authentication, comprehensive access controls, application security and end-to-end encryption. However, these measures may not be sufficient. Appropriate risk assessment should be conducted to understand risks that may present in the business environment.

Embedding AI onto the design not only increases efficiencies but also the quality of the processes. The use of robotics controlled by AI in the automation process is capable of sending data to the cloud. However, if the integrity of AI modules is compromised, it will affect the processes and eventually the end-product particularly when RPA is implemented in a smart manufacturing environment.

Big Data

Big data has characteristics of a variety of data types and sources, velocity, volatility, and value to name a few, which introduces risks. Data variety leads to challenges and complexity in data management which subsequently introduces vulnerabilities. The volatility of data can cause challenges in keeping the audit trail and in managing it. The usage of smart devices has led to the increase of data exponentially. With the IR4 in the pipeline, more data is to be collected in various forms from different sources of both digital and physical environments. This data is generated by traditional means of digital platforms and through sensors and IoT devices embedded in the physical environment such as critical infrastructures and smart cities.

Due to this, its collection and processing face challenges, as it is not inherent in the normal digital setting. Chaudhari and Srivastava (2017) argue that with the huge amount of data using current technologies and efficient cryptography algorithms, the processing to encrypt database is slow; thus affecting the data processing time and results. It is very challenging to segregate data based on its importance and determine its ownership when the volume is massive (Chaudhari & Srivastava, 2017). Data is protected in accordance with its classification. The higher its classification level, the higher the budget required to protect them are. Thus, it is unnecessary to implement countermeasures when the data does not need one.

Cybersecurity Concerns

The increasing interdependencies on technologies and cyberspace have caused users' adoption of technologies but it is not aligned with the security counter-measures in place, particularly with the usage of advanced technologies as in IR4.0. The issue is that there is an insufficient workforce that has relevant capabilities to deal with the security of the technologies. It is estimated that the cost of cybercrime globally will reach USD 10.5 trillion annually by the year 2025 (Morgan, 2020). The following sub-sections discuss the security concerns in manufacturing, agriculture and services.

Manufacturing

The evolution of digital technologies has changed the way manufacturing plants operate. IR4.0 has led to increased smart manufacturing where conventional manufacturing which typically uses industrial control system (ICS) is converged with the production line using IP communication. Smart manufacturing or also referred to as Industry 4.0 provides the ecosystem with advanced technology architecture such as IoT, robotic and AI. Naturally, the manufacturing environment provides the usage of OT and IT systems. ICS which is part of OT is used not only in the manufacturing environment but also in other environments such as in power plant, water, petroleum and nuclear plant. The exploitation of OT systems does not only affect the integrity of safety systems but can cause physical effects (Copic et al., 2019) that may implicate the public.

In a manufacturing plant, the production line may start using machine-to-machine, IoTs and sensors for data collection purposes. However, increasing interconnectivity due to interconnected devices in the manufacturing plants has increased security risks, which can pave the way for cyber perpetrators to exploit any vulnerability in the plant's environment.

This is due to ICS being designed as a closed system and developed without built-in security (Mirian et al., 2016). Previously, the manufacturing environment was secure due to its physical access restriction where they are isolated from other parts of the system. But when systems are interconnected with devices and connected to a wider corporate network that provides a gateway to the public network, the risks of being exposed increased, thus, making the network environment less secure.

With IR4.0 deployment, manufacturers are exposing themselves to new technologies in an interconnected environment (Copic et al., 2019). The complexities are intensified with the adoption of advanced technologies which the environment has to be customized to. Shifting from mass production to mass customization may require the integration of new technologies with legacy systems that may expose security weaknesses in various parts of the ecosystem.

The need for monitoring and controlling processes in an ICS requires sensors and devices to allow data to be acquired and transmitted from the sensors to the Supervisory Control and Data Acquisition (SCADA) where the monitoring of the systems will be performed. In increasing the production performance, sensors of the plants are sent directly to the cloud.

ICS in the OT environment has a lack of security features (Mirian et al., 2016). Increasing sensors and IoT devices in the IR4.0 technology architecture will intensify the complexity of the environment. Thus, it is imperative to establish a dedicated security team to understand the architecture and the communication of various protocols and components in this environment. Using robotics that automatically transmit data to the cloud platform can elevate further risks if communications are not configured securely.

In making a better-controlled environment, the usage of IoT and sensors that gather different types of data (i.e., humidity, temperature and carbon dioxide) in real-time is part of the whole system architecture. Thus, data accuracy is critical and the integrity and confidentiality of collected data should not be compromised at any cost.

IR4.0 provides a platform that allows manufacturing processes to be operated in an automated manner in a hyperconnected environment through smart components and devices that are connected to the cloud platform. The challenges will be increased with the adoption of a 5G environment that allows more IoT devices to be deployed since it provides a faster network with low latency. This smart manufacturing environment requires smart devices that have sensor abilities for automation in processes. If they are not properly secured, they will be exposed to vulnerabilities.

In providing communication on TCP/IP platform, the IEC-60870-5-104 standard protocol used in the SCADA system (which is a subset of the ICS system) can be exploited; thus, eventually disrupting the normal operation of critical infrastructures. In an experiment conducted by Qassim et al. (2018), insufficient security protection and attack prevention mechanisms at the application and data link layers of this protocol have an impact on the implementation of SCADA systems. Based on this evidence, similar systems using the same protocol including those of manufacturers can experience the same security risk.

Agriculture

The deployment of sensors in the agricultural sector in gathering and transmitting data through wireless sensor networks using secure mechanisms is nothing new in many countries. Despite the agriculture sector being able to obtain insights for better farming through data-driven technologies, Malaysia faces many challenges in adopting digital technologies. Apart from the lack of a reliable internet connectivity in rural areas, skilled workers who are able to interpret and make use of the data are scarce (Tompkins, 2020).

In responding to IR4, relevant technological skill sets such as those pertaining to IoT, AI and big data analytic capabilities are important to be developed in deploying the technologies and interpreting the data collected. Data collection and monitoring for the agriculture sector is very specific to its geographical location, weather, and climate for analytical purposes to get the best results for productions. Other data collected includes data to understand infrastructure or transportation, as well as storage and market proximity (Prodanović et al., 2020). Security threats in agriculture are very much similar to other sectors where various mechanisms were deployed to address malicious adversaries.

A security model developed by Prodanović et al. (2020) that is independent of network topology and infrastructure provides data protection from sensors to wireless networks using encryption capabilities such as public key infrastructure (PKI). In the facilitations of various data sources for remote access, satellite functionalities are used. Although this model is able to fulfil data protection in its ecosystem in an efficient manner, the drawback is that it increases energy consumption due to the frequent exchanges of the secret key during the encryption process between the receiver and sender.

Services

The increase of data-driven services such as digital education, digital trade and e-commerce require a large pool of data transmitted across borders (Aaronson, 2019). E-commerce activities involve the transmission of personal data with credit card details for a transaction to take place. Customers risk their data when purchasing goods over the internet as there is no assurance that security measures have been adopted in their infrastructure. Digital trade is defined as the digital transactions of goods and services involving cross-border transactions which are digitally ordered, facilitated or delivered (OECD-IMF, 2018), where three parties are heavily participating; consumers, firms and governments. On the other hand, e-commerce is categorized as a type of digital trade (ESCAP, 2021).

Apart from security risks, there are also great concerns about digital trade and cross-border data which are expected to grow exponentially with the increasing use of mobile devices and usage of social media. E-commerce activities and business transactions are now increasingly performed on the go. This has also raised concerns by how companies with big names regulate these services and their associated data. The use of public and commercial cloud computing services has created more cross border challenges for some jurisdictions (Abed & Chavan, 2019).

Cybersecurity has been identified as one of the most significant challenges for online communities to use digital technology platforms including for e-commerce activities. Bartczak (2021) argues that the digital platform can be an obstacle to the effectiveness of e-commerce due to security threats. The digital platform which enables buyers and suppliers to trade requires trust for the exchange of product (from supplier) with money and personal data (from customer). Thus, security measures that enable this trust and confidence should be provided while the transaction occurs while the personally identifiable information (PII) should be well-protected. Personal data analysis is crucial to understanding purchasing patterns of individuals for marketing purposes. However, not ensuring privacy measures are in place can cause problems for businesses (Venturini & Rogers, 2019).

Other Security Concerns

Apart from existing features highlighted in previous sections, there are other aspects such as 5G technologies and legacy systems that are worth considering due to the implications of their deployment.

Transition to 5G

The current infrastructure of the 4G network is already overloaded with subscribers' usage of communication and data transmission, particularly in major cities, where the population density is high. This has caused regular slowdowns of data rates transmission occurring commonly during peak hours that demand larger network capacity. On the contrary, 5G will be able to better meet the new demands and needs of faster connectivity to allow massive data transmission from one point to many points, whilst at the same time allowing massive data processing at the other ends of the destination, either as devices or systems. Due to its hyper network connectivity, faster-speed transmission and low latency (Kubota, 2020), 5G technology will be a critical enabler of new business services and products that will drive companies forward in terms of efficiency and innovation.

5G has a direct influence on IR4.0 ecosystem including cloud computing, IoT, robotics and automation processes that trigger better manufacturing routines, telemedicine and the evolution of transportation industries. As the industries are moving towards IR4.0, automated systems and devices via AI will be preferred, where data is transmitted from one point to many destination points instead of on a one-way traffic. Kubota (2020) posits that the adoption of 5G networks will be dominated by IoT devices instead of consumers. This indicated autonomous devices will be dominating the network. However, the number of cyber-attacks is increasing on these devices due to poor protection.

One of the main purposes of 5G technology development is to resolve cybersecurity issues known to its predecessors. Thus, 5G has incorporated security controls that address many cybersecurity threats faced in the previous infrastructure networks. These controls include new mutual authentication capabilities, enhanced subscriber identity protection and other additional security mechanisms in the communication protocols (Kubota, 2020). However, there are still security concerns in 5G that were raised by security researchers. Substantial known security threats in the 5G ecosystem can affect the 5G telecommunication infrastructure due to improper implementation as it can open more potential entry points to hackers.

Legacy System

The continuous usage of legacy systems makes the environment more complex as there is less support and available technical expertise to work on the programming code of these systems. Legacy systems are known for the absence of security features. Even if the systems have such security features, they are no longer supported by vendors. Despite being used in isolation, which is a popular approach in ensuring legacy systems are protected, there are potential security risks when removable disks such as USB flash drives are used as a medium of data transfer within the environment.

In transforming into Industry 4.0, older systems which are not sophisticated can be a barrier where these systems are unable to handle huge data processing. This could lead to the unavailability of services (Tay et al., 2021). Not only that, it can also pose security risks that are not only caused by the new technologies being implemented but also due to the interactions between new and legacy systems (Copic et al., 2019). Organizations have their own reasons for keeping their legacy systems. Apart from a massive cost that can be incurred from moving into new platforms where new technologies are readily available for adoption (e.g. cloud and IoT), there are other aspects that hinder the transformation including a shortage of skilled workers that are capable in translating the old programming code.

Although legacy owners tend to move into modernization that integrates the legacy systems with new technologies, the main challenge is to ensure the integrity, confidentiality and available data is not compromised as the integration process takes place. However, there has been a lack of framework in addressing such security concerns despite the modernization processes that have taken place. Frequently, security aspects are due to oversights that contribute to vulnerabilities against cybercrimes.

Addressing Cybersecurity Concerns

A holistic mechanism is crucial in approaching the issue to address the gap effectively. This comprises three main pillars in addressing cybersecurity namely, people, process and technology. Due to the nature of cybersecurity as a public good, cooperation can be the salient element in binding the pillars with relevant stakeholders and regulatory bodies to assure that required steps are taken to protect the digitized ecosystem (Daud, 2017). The increase of connectivity and interdependencies along with new technologies demand a holistic approach to increase functions and efficiency in organizations and businesses. In order to have a secure IR4 environment, security countermeasures should be considered at various levels, including both national and organizational levels.

Cybersecurity Readiness in IR4.0

The readiness of organizations in their cybersecurity measures provides an indicator on how organizations can overcome security challenges in their business environment particularly when organizations are very dependent on ICT and internet connection. Similarly, security researchers have emphasized significant security challenges for organizations to embrace IR4 (Tay et al., 2021). Berlilana et al. (2021) asserts how cybersecurity readiness affects the performance of organizations. A pilot study amongst 100 technology companies in Malaysia show their readiness to adopt Industry 4.0 and that they have the necessary knowledge to embrace it (Soomro et al., 2021). However, security was not a part of the study. In comparison to a study conducted by Tay et al. (2021) amongst SMEs in Malaysia, the findings conclude that Malaysian manufacturing companies are at its early stage of transforming from existing manufacturing to Industry 4.0. This is due to challenges faced by the businesses including lack of technology, competent and knowledgeable workers, as well as the need to change mindsets. Another challenge faced in adopting Industry 4.0 is data management and integration due to increasing data that is and needs to be collected.

Technology readiness is another challenge companies face in embracing IR4.0, not only due to the high-capital investment that is required to provide the technologies, but also the complex integration necessary in transforming existing manufacturing into smart manufacturing. This also requires skilled workers with the right talent which are in short supply (Tay et al., 2021). This is supported by Javaid et al. (2021) where one of technological challenges arose from the integration process of IIoT connectivity, where it is not cost effective for transformation to take place in small scale industries. This finding is important since technology readiness is directly related to security readiness (Berlilana et al., 2021).

Increasing of platforms through expansion of technology infrastructure with different protocols for data communication can lead to security and privacy issues when skilled workers are not available to perform the transformation process. Incompatibility of protocols and improper use of application programming interfaces (APIs) can lead to unnecessary data exposure. In the case of the Facebook and Cambridge Analytica scandal, prior to changing the API in restricting data access, data profiles from the social media (Facebook) were retrieved and accessed by other applications which had implications on the US presidential campaign in 2018 (Venturini & Rogers, 2019). Having legacy systems in place will complicate things although enterprises have their reasons for keeping their legacy systems. Apart from cost-effectiveness, a lack of those who are able to deal with old programming codes and systems can be challenging. Although cybersecurity is not considered one of the critical challenges, data management issues can eventually lead to security implications if not tackled prudently (Tay et al., 2021).

Thus, a central monitoring programme can be developed to gauge the level of security readiness in facing cybersecurity threats when embracing IR4. The security-readiness can be measured from the readiness of people, processes and technology in managing the security of the equipment and environment. Identifying the gaps will help organizations to focus on areas they are lacking; thus, necessary funds can be allocated, and specific programmes can be driven to narrow the gap.

Developing relevant skills is important to embrace the future workforce in IR4. In realizing the IR4 agenda, technical and vocational education and training (TVET) have been identified as an important factor where the Ministry of Human Resources and the Ministry of Education Malaysia have developed a policy for implementing the TVET agenda (Mohd Ishar et al., 2020). Similarly, the National IR4.0 policy also put forward the agenda for developing talents to address current and future needs in IR4. The role of the Human Resources Development Fund has been identified to drive the initiative (Economic Planning Unit, 2021b).

In addition to this, through pillar four of the MCSS, Malaysia Digital Economy Corporation (MDEC) will spearhead initiatives to increase skilled workers in cybersecurity, while CyberSecurity Malaysia through Global Accredited Cybersecurity Education (ACE) has implemented people certification initiatives associated with certain security domains (National Security Council, 2020). In preparation to mitigate cybersecurity risks, training and educating of future knowledgeable workers means instilling cybersecurity education at the primary, secondary and tertiary education levels.

Security by Design

Due to its complexities, organizations that wish to move to IR4 should be establishing a security architecture framework which ensures that the infrastructure the organization will be operating on is safe. Frequently, security is implemented as an afterthought. Rightfully, it should be embedded at the design stage of the systems' development right from the beginning. The design should incorporate as security requirements and these should be defined and incorporated as part of the procurement process. Getting vendors to develop systems with security features right from the beginning of the systems' development cycle is more cost-effective than fixing security issues when development has been completed.

For instance, for organizations that consider deploying RPA technology, the nature of RPA should be well understood. By understanding RPA in depths, its risks can be minimized. Since RPA is a type of software, its security threats are no different from other software. For instance, organizations should be considering security controls that are related to software such as multi-factor authentication, a comprehensive access control, application security, and end-to-end encryption when deploying RPA (Gautam, 2022). It is evident that the way RPA is implemented requires a secure environment it can operate. Less privilege access should be allowed for bots to perform their assigned tasks (Kosi, 2019). Bad programming can also cause the systems' resources to be over consumed and this could lead to the unavailability of services.

In the context of huge data being processed, understanding the effects of security practices for businesses needs to be understood. For example, huge data can cause systems to become slower while being processed, not to mention implementing encryption on the database to protect data confidentiality could do the same as well. Instead of encrypting the whole database that can slow down the processing time, Chaudhari and Srivastava, (2017) suggest systematically encrypting individual parts of the data to allow a certain level of flexibility in the processing. However, this approach can cause security to be compromised.

Companies should ensure that when producers implement cryptographic algorithm on products and services, reference is made to the list of National Trusted Cryptographic Algorithm that is based on existing cryptography algorithms (CyberSecurity Malaysia, 2020). This is the list of trusted existing cryptographic algorithms for implementation that has gone through evaluation and testing by cryptographic experts in Malaysia.

Organizations should produce the security requirements for any systems to be developed apart from the system's functionalities. These requirements should be embedded as part of the procurement process as this is a more cost-effective approach than trying to fix the systems after bugs are found when a vulnerabilities assessment is performed.

Structured Security Processes

This section looks at several available structured security processes. Information security is defined as the preservation of the availability and integrity of information regardless if the information is in the digital or non-digital format (International Standard Organisation, 2013). Cybersecurity is defined as the safeguarding of society, people, organizations and nations from cyber risks and keeping cyber risk at a tolerable level (International Standard Organisation, 2020).

Security is a process. Due to the interconnectivity and interdependencies in an IR4.0 environment, security should be managed in a comprehensive approach that is proactive and responsive (Daud, 2017). The proactive domain requires security processes to be implemented with the intention to prevent security breaches which includes having risk assessment and security assurance. This is discussed further in the next sub-section. Although in the responsive domain, an incident management is invoked upon security incidents occurring, relevant plan should be developed and tested.

The information security management system (ISMS) which is based on ISO/IEC 27001:2013 is considered the de facto standard in managing security in organizations as its main objective is to protect information assets. It provides a holistic approach which is risk-centric. It is a wise approach for organizations to perform a risk assessment before embarking on new systems or technologies deployment, thus prudent critical decisions can be made. Security is best managed by implementing security controls derived from a security risk assessment exercise.

Security risk assessment is considered the basis for making decisions where security controls can be identified much earlier before performing the cost-benefit analysis. For instance, organizations frequently utilize cloud storage for cost-efficiency purposes. One of the important critical decisions is the identification of security controls to protect sensitive data such as personal data in the cloud environment. Without conducting a proper risk assessment, the risk in of security breaches happening eventually defeats the whole purpose of deployment. In understanding the risks that organizations face, risk assessment is a brilliant approach to decide what security countermeasures should be implemented while minimizing any risks that materialize.

Several standards are related to risk management such as ISO/IEC31010 and ISO/IEC 27005. The former discusses all types of risks while the latter is specifically developed to manage risks in information security. In a manufacturing environment, the quality of implemented ISMS is significant to ensure the security of IT and OT in the manufacturing industry is monitored and continually improved as the adoption of IEC 62443 standard used in the industry is based on the ISMS series (Kosmowski et al., 2019). For industries that deploy ICS, relevant standards and guidelines that are typically used to secure ICS operating environment include ISO 27019, IEC/ISA 62,443 and NIST SP 800–82 (Maesschalck et al., 2022).

Apart from ISMS, another important security process is the business continuity management (BCM). Not only it is capable of increasing resiliency in organizations, but it also provides data recovery strategies as the continuity plan is developed. Since data is the lifeblood of businesses, assuring that data and information are available in the event of a crisis or disaster is fundamental. Responding to security incidents should always start with consistent performing of security logs analysis that monitors security events. Understanding patterns means being able to detect impending attacks. When the security threshold is met, the incident management team, through its process, will handle such incidents and assume recovery processes. Through this process, businesses are able to identify if data recovery is required. Thus, having a tested business continuity plan in place can be a significant safeguard.

Additionally, organizations may also need its workforce to have some data forensic capabilities to ensure data collected or acquired from the systems upon security incidents are preserved in accordance with international best practices; thus, to be acceptable in any court of law. Since an IR4 ecosystem provides a variety of data sources, understanding the right method to identify infected devices and collect data is crucial. So is the ability to provide relevant analysis in understanding the root cause of security breaches. This provides the lessons learnt to prevent future security breaches.

Security Assurance on Products and Platforms

Security assurance on products and services is significant to generate trust and confidence in users who are using them. IR4.0 device manufacturers are expected to implement security functionalities in their products that will enable security to function appropriately (ENISA, 2019). Devices should allow the systems to be updated accordingly. This feature is important as one of the vulnerabilities that can be easily exploited is when devices are not patched or updated to manage known vulnerabilities. Device manufacturers that claim their products are secure should be sending their products for security evaluation performed by accredited laboratories. This is supported by Tay et al. (2021), wherein a manufacturing environment, there are always risks when compromised devices are connected to the network. These devices may not seem to be vulnerable at one point, but they might be exploited later in the future. Thus, the responsibility of connected devices should not only fall on the users of the device but also should be shared with the device manufacturers.

In relation to this concern, it is imperative for device manufacturer or developers to go through product security evaluation process and later also go through a certification process. Product certification that is internationally acknowledged is a Common Criteria which is based on the international ISO/IEC 15408 scheme. This evaluation covers all aspects of security including systems, databases, applications, devices, as well as network etc.

For e-commerce entities, having a certification that signifies them as a trusted website is imperative to demonstrate the trustworthiness of their business platform over the internet. Web application attacks can cause data to be stolen from the database through cross-site scripting or SQL injection; thus, can cause a business to lose market shares due to the loss of reputation and trust. In building global trust and confidence for businesses on an e-commerce platform, World Trustmark and Trade Alliance through its country representatives offers a mechanism to validate that a website is secure as well as conform with good practices and standards.

Although the transition to 5G is gradual, the stakeholders in the ecosystem should be prepared. The distinct characteristic of 5G is "network slicing", a software-based prominent feature that allows virtual network implementation with the aim to create subnets in meeting specific needs. Security evaluation and testing of this platform and 5G telecommunication devices and applications are imperative to provide trust and assurance of the network. As part of the preparation towards a 5G security ecosystem, Malaysia through CyberSecurity Malaysia has established 5G security test lab to meet this purpose (Bernama, 2021).

Identifying security threats and managing security vulnerabilities are important. A consistent vulnerabilities management programme is fundamental to detect security vulnerabilities and avoid them from being exploited by performing periodic vulnerabilities and penetration testing (VAPT). Any changes in security configuration and implementation should be appropriately documented to trace problem when security events occur. Apart from assessing infrastructures such as servers, hosts and security devices, web applications and databases should not be left out.

Due to the increasing use of software, source code review is another area to be looked at. This is critical to ensure that there are no coding weaknesses such as hard-coded password or a backdoor left opened by the developer. As for entities that use cloud platform as part of their business operation, going through a cloud security audit is wise to ensure that their data hosted on the cloud are secure and protected. Using the ISO/IEC 27017 as a guide, cloud subscribers are able to implement security controls which are based on the security catalogue listed in ISO/IEC 27002 (International Standard Organisation, 2015).

Skilled Workers, Talents and Awareness

Creating talented and skilled workers with a diverse skill set has been identified as an urgent need for the manufacturing sector and in e-commerce (Ministry of International Trade and Industry, 2018). This is particularly challenging as businesses learn how challenging it can be for employees to respond to security incidents. Even hiring and retaining these cybersecurity talents can be difficult and that can be a major threat to businesses (World Economic Forum, 2022). Thus, relevant programmes should be established focusing on developing new skills, up-skilling and re-skilling to increase workers with cybersecurity knowledge. Although some entities are comfortable with the adoption, the lack of readiness due to shortage of skilled workforce should not hinder businesses particularly SMEs to adopt IR4 technologies (Tay et al., 2021). This is to prevent a wider gap in the adoption that can cause Malaysia to be left behind in the world economy.

For organizations to plan on ensuring workers with skills and competencies, it has to be prioritized and built around the business focus. Advanced technologies such as RPA, AI and 5G are software driven; thus, competencies and relevant skills on software security should be enhanced in preparation to adopt these technologies.

While manufacturers have been adopting IT with less emphasis on security in previous manufacturing environment, it is timely for them to establish a dedicated security team that will oversee the security implementation, integration, and monitoring of various technologies. Since cybersecurity is everyone's responsibilities, a comprehensive awareness programmes should be developed targeting various categories of audiences depending on their role in the organizations. Not only for those who will be responsible in handling security, but also for typical IT and internet users in organizations. Ransomware through social engineering attack still dominates the list of attacks regardless of the industries and this demonstrates the weaknesses of users in dealing with daily operational activities (World Economic Forum, 2022).

Security Governance

IR4.0 involves various owners and stakeholders through several national policies that were established in addressing IR4.0 initiatives for the country. Security aspects were addressed through these policies in different platforms. Thus, a centralized platform or an agency with a regulatory role in cybersecurity can be seen as an approach that is able to govern and monitor the implementation of these security initiatives in a more effective manner.

Cybersecurity can be a significant contributing factor to the success of the IR4.0 implementation in Malaysia. With the 5G deployment in Malaysia, it is anticipated that there will be an exponential growth of mobile devices, sensors or IoT devices usage in business operations as well as the public at large. However, without security countermeasures and strategies in place, the massive deployment of these devices can be turned into a platform for possible attacks to take place. Thus, the regulatory role is crucial to secure devices to be used, as well as the platform and environment within the ecosystem for a safe and secure computing.

Since the spread of IR4.0 through the national economy requires massive data collection, it is possible for regulatory bodies to effectively govern information exchange and operations, particularly when it involves cross-border data transmission related to personal data. Data protection should be strengthened where data collection from personal mobile devices and heavy usage of social media is observed. Cross-border data flows should be governed by policy makers by considering approaches at both national and international levels.

At the national level, Malaysia through the Personal Data Protection Act (PDPA) has delineated personal data between the commercial and non-commercial entities. Although data privacy protection is regulated for commercial entities, it is not the same with government data. It is important to note that trade data does not only involve commercial data. At international levels, policy makers should be heavily involved in public forums and also make collaborative efforts to have the voices of people in the country be heard and for citizens to have better control over their own data (Aaronson, 2019).

Conclusion

In realizing the nation's aspirations, not only does the government and intermediary organizations play an important role to support the development of the digital infrastructure and regulate its activities, but they also need to ensure that the online platforms are secure for all stakeholders to serve their intended purposes. IR4.0 is a mechanism that encompasses various technologies with a blend of advanced technologies and legacy systems, which require several shared operations from strong embedding ecosystem support. These technologies require security measures and strategies on the massive data produced, and hence, a new paradigm of collective and collaborative coordination between users and producers is needed.

To achieve such a harmonious network of connectivity and coordination, security readiness of IR4.0 components in relevant industries and sectors need to be thoroughly understood. For businesses to embark into this complex platform, the workforce who will be implementing, integrating and monitoring the systems should be well equipped with relevant skills, knowledge and expertise. Thus, the need of a workforce in the cybersecurity domain should not be taken lightly. To fully benefit from the transformation in an effective manner, having security features embedded in every part of systems and processes in IR4.0 ecosystem right from the beginning is imperative, and should not be an afterthought. There is also a need for co-operation among all national cybersecurity organizations to strengthen protection and privacy of users and producers as cyber-attacks are borderless.

References

Aaronson, S. A. (2019). Data Is Different, and that's Why the World Needs a New Approach to Governing Cross-border Data Flows. *Digital Policy, Regulation and Governance, 21*(5), 441–460. https://doi.org/10.1108/DPRG-03-2019-0021

Abed, Y., & Chavan, M. (2019). The Challenges of Institutional Distance: Data Privacy Issues in Cloud Computing. *Science, Technology and Society, 24*(1), 161–181. https://doi.org/10.1177/0971721818806088

Bartczak, K. (2021). Cybersecurity as the Main Challenge to the Effective Use of Digital Technology Platforms in E-Commerce. *European Research Studies Journal, XXIV*(Issue 2B), 240–256. https://doi.org/10.35808/ersj/2230

Berlilana, Noparumpa, T., Ruangkanjanases, A., Hariguna, T., & Sarmini. (2021). Organization Benefit as an Outcome of Organizational Security Adoption: The Role of Cyber Security Readiness and Technology Readiness. *Sustainability (Switzerland)*, 13761, https://doi.org/10.3390/su132413761

Bernama. (2021). *My5G Test Lab To Strengthen Malaysia's Cyber Security Capabilities, Says Minister.* https://www.kkmm.gov.my/en/public/news/21019-my5g-test-lab-to-strengthen-malaysia-s-cyber-security-capabilities-says-minister

Chaudhari, N., & Srivastava, S. (2017). Big Data Security Issues and Challenges. *Proceeding - IEEE International Conference on Computing, Communication and Automation, ICCCA 2016, May*, 60–64. https://doi.org/10.1109/CCAA.2016.7813690

Copic, J., Leverett, É., Quantrill, K., Cameron, P., Douglas, T., Bourdeau, J., & Evan, T. (2019). *Managing cyber risk in the Fourth Industrial Revolution: Characterising Cyber Threats, Vulnerabilities and Potential Losses. July*, 24.

CyberSecurity Malaysia (2020). *Existing Cryptographic Algorithm for MySEAL (AKSA MySEAL).* https://myseal.cybersecurity.my/en/aksa.html

Daud, M. (2017). *National Information Infrastructure Organisations and Cyber Security Compliance in Malaysia. November*, 1–186. http://studentsrepo.um.edu.my/8959/9/maslina.pdf

Dror, N. (2022). *Top 5 Security Threats Facing Artificial Intelligence and Machine Learning.* https://hubsecurity.com/blog/cyber-security/security-threats-for-ai-and-machine-learning/

Economic Planning Unit (2021a). Malaysia Digital Economy Blueprint. In *Economic Planning Unit Prime Minister's Department.* https://www.epu.gov.my/sites/default/files/2021-02/malaysia-digital-economy-blueprint.pdf

Economic Planning Unit. (2021b). *National 4IR Policy.* https://www.epu.gov.my/sites/default/files/2021-07/National-4IR-Policy.pdf

ENISA (2019). *Enisa Lists High-Level Recommendations To Different Stakeholder Groups in Order To Promote Industry 4.0 Cybersecurity and Facilitate Wider Take-Up of Relevant Innovations in a Secure Manner. 2 Industry 4.0 Cybersecurity: Challenges & Recommendations.* https://www.enisa.europa.eu/publications/industry-4-0-cybersecurity-challenges-and-recommendations

ESCAP (2021). *Regional Digital Trade Integration Index Guidelines.* https://hdl.handle.net/20.500.12870/4153

Gautam, R. (2022). *4 Security Must-Haves for a Safe RPA Solution A secure, Quality Platform.* https://www.automationanywhere.com/company/blog/product-insights/four-security-must-haves-for-a-safe-rpa-solution

International Standard Organisation (2013). ISO/IEC 27001:2013 Information Technology - Security Techniques - Information Security Management Systems - Requirements. In *International Standard* (Issue 19, p. 30), downloaded on November 30, 2022 from https://www.iso.org/standard/54534.html

International Standard Organisation (2015) Information technology — Security techniques — Code of practice for *Information Security Controls Based on ISO/IEC 27002 for Cloud Services*, downloaded on November 30, 2022 from https://www.iso.org/standard/43757.html

International Standard Organisation (2020). *ISO/IEC TS 27100:2020 Information Technology Cybersecurity — Overview and Concepts*, downloaded on November 30, 2022 from https://www.iso.org/obp/ui/#iso:std:iso-iec:ts:27100:ed-1:v1:en

Javaid, M., Abid Haleem, Pratap Singh, R., Rab, S., & Suman, R. (2021). Upgrading the Manufacturing Sector via Applications of Industrial Internet of Things (IIoT). *Sensors International,* 2(August), 100129. https://doi.org/10.1016/j.sintl.2021.100129

Kosi, F. (2019). *Robotic Process Automation (RPA) and Security,* Masters Thesis, Mercy College, Bronx Campus.

Kosmowski, K. T., Śliwiński, M., & Piesik, J. (2019). Integrated Functional Safety and Cybersecurity. Analysis Method for Smart Manufacturing Systems. *Task Quarterly,* 23(2), 177–207. https://doi.org/10.17466/tq2019/23.2/c

Krebs, B. (2016). *Who Makes the IoT Things Under Attack?* https://krebsonsecurity.com/2016/10/who-makes-the-iot-things-under-attack/

Kubota, A. (2020). 5G Security. *IEICE Communications Society Magazine,* 14(3), 254–260. https://doi.org/10.1587/bplus.14.254

Kuzlu, M., Fair, C., & Guler, O. (2021). Role of Artificial Intelligence in the Internet of Things (IoT) Cybersecurity. *Discover Internet of Things,* 1(1). https://doi.org/10.1007/s43926-020-00001-4

Maesschalck, S., Giotsas, V., Green, B., & Race, N. (2022). Don't Get Stung, Cover your ICS in Honey: How do Honeypots Fit within Industrial Control System Security. *Computers and Security,* 114, 102598. https://doi.org/10.1016/j.cose.2021.102598

Ministry of International Trade and Industry (2018). Industry 4WRD: National Policy on Industry 4.0. In *Ministry of International Trade and Industry.* https://www.miti.gov.my/miti/resources/National Policy on Industry 4.0/Industry4WRD_Final.pdf

Mirian, A., Ma, Z., Adrian, D., Tischer, M., Chuenchujit, T., Yardley, T., Berthier, R., Mason, J., Durumeric, Z., Halderman, J. A., & Bailey, M. (2016). An Internet-wide View of ICS Devices. *2016 14th Annual Conference on Privacy, Security and Trust, PST 2016,* 96–103. https://doi.org/10.1109/PST.2016.7906943

Mohd Ishar, M. I., Wan Derahman, W. M. F., & Kamin, Y. (2020). Practices and Planning of Ministries and Institutions of Technical and Vocational Educational Training (TVET) in Facing the Industrial Revolution 4.0 (IR4.0). *Malaysian Journal of Social Sciences and Humanities (MJSSH),* 5(3), 47–50. https://doi.org/10.47405/mjssh.v5i3.374

Morgan, S. (2020). Cybercrime To Cost The World $10.5 Trillion Annually By 2025. *Cybersecurity Ventures,* 1. https://cybersecurityventures.com/cybercrime-damages-6-trillion-by-2021/%0A;https://cybersecurityventures.com/hackerpocalypse-cybercrime-report-2016/

MyCert (2022). *Getting IT and Digital Information.* accessed on June 17 from https://www.malaysia.gov.my/portal/content/30870

National Security Council (2020). *Malaysia CyberSecurity Strategy 2020–2024,* downloaded on July 20, 2022 from https://asset.mkn.gov.my/wp-content/uploads/2020/10/MalaysiaCyberSecurityStrategy2020-2024.pdf

Nematollahi, M. R., & Khorasani, K. (n.d.). *Note For National Defence: Artificial Intelligence: Cybersecurity Challenges,* MINDS, downloaded on November 30, 2022 from https://www.concordia.ca/content/dam/ginacody/research/spnet/Documents/BriefingNotes/SpaceandCyberspace/BN-93-Space-and-cyberspace-Nov2021.pdf

Obaidat, M. A., Obeidat, S., Holst, J., Hayajneh, A. Al, & Brown, J. (2020). A Comprehensive and Systematic Survey on the Internet of Things: Security and Privacy Challenges, Security Frameworks, Enabling Technologies, Threats, Vulnerabilities and Countermeasures. *Computers,* 9(2), 44. https://doi.org/10.3390/computers9020044

OECD-IMF (2018). Towards a Handbook on Measuring Digital Trade. *Thirty-First Meeting of the IMF Committee on Balance of Payments Statistics*, Washington DC: the OECD and the IMF, downloaded on August 12, 2022 from https://www.imf.org/external/pubs/ft/bop/2018/pdf/18-07.pdf

Prodanović, R., Rančić, D., Vulić, I., Zorić, N., Bogićević, D., Ostojić, G., Sarang, S., & Stankovski, S. (2020). Wireless Sensor Network in Agriculture: Model of Cyber Security. *Sensors (Switzerland)*, *20*(23), 1–22. https://doi.org/10.3390/s20236747

Qassim, Q. S., Jamil, N., Daud, M., Ja'affar, N., Yussof, S., Ismail, R., & Kamarulzaman, W. A. W. (2018). Simulating Command Injection Attacks on IEC 60870-5-104 Protocol in SCADA System. *International Journal of Engineering and Technology(UAE)*, *7*(2.14 Special Issue 14), 153–159. https://doi.org/10.14419/ijet.v7i2.14.12816

Rasiah, R. (2019). Building Networks to Harness Innovation Synergies: Towards an Open Systems Approach to Sustainable Development. *Journal of Open Innovation: Technology, Markets and Complexity*, 5, 70

Sinha, S. (2021). *State of IoT 2021: Number of Connected IoT Devices Growing 9% to 12.3 B*. IoT Analytics. https://iot-analytics.com/number-connected-iot-devices/%0Ahttps://iot-analytics.com/number-connected-iot-devices/?utm_source=IoT+Analytics+Master+People+List&utm_campaign=8de8177e37-Cellular+IoT+LPWA+Tracker+Blog&utm_medium=email&utm_term=0_3069fbcae4-8de

Soomro, M. A., Hizam-Hanafiah, M., Abdullah, N. L., Ali, M. H., & Jusoh, M. S. (2021). Industry 4.0 Readiness of Technology Companies: A Pilot Study from Malaysia. *Administrative Sciences*, *11*(2), 56. https://doi.org/10.3390/admsci11020056

Tay, S. I., Alipal, J., & Lee, T. C. (2021). Industry 4.0: Current Practice and Challenges in Malaysian Manufacturing Firms. *Technology in Society*, *67*(September), 101749. https://doi.org/10.1016/j.techsoc.2021.101749

Tompkins, S. (2020). Getting Ready for Agriculture 4.0. *The Star*. https://www.thestar.com.my/opinion/letters/2020/07/16/getting-ready-for-agriculture-40

Venturini, T., & Rogers, R. (2019). "API-Based Research" or How can Digital Sociology and Journalism Studies Learn from the Facebook and Cambridge Analytica Data Breach. *Digital Journalism*, *7*(4), 532–540. https://doi.org/10.1080/21670811.2019.1591927

World Economic Forum (2022). *Global Cybersecurity Outlook 2022*. https://www3.weforum.org/docs/WEF_Global_Cybersecurity_Outlook_2022.pdf

Index

Note: **Bold** page numbers refer to tables and *italic* page numbers refer to figures.

Printed in the United States
by Baker & Taylor Publisher Services